Francis C. (Francis Cruger) Moore

Water Works and Pipe Distribution

Francis C. (Francis Cruger) Moore

Water Works and Pipe Distribution

ISBN/EAN: 9783744649964

Printed in Europe, USA, Canada, Australia, Japan

Cover: Foto ©berggeist007 / pixelio.de

More available books at **www.hansebooks.com**

MECHANICAL ENGINEERING

FOR BEGINNERS.

BY

R. S. M'LAREN.

With Numerous Illustrations.

LONDON:
CHARLES GRIFFIN & COMPANY, LIMITED;
EXETER STREET, STRAND.
1908.

PREFACE.

OF books dealing with the various branches of Mechanical Engineering there is an immense choice. Reliable textbooks can be found dealing with almost any subject upon which an engineer can desire information; but when asked by a beginner, say by a youth whose friends have just placed him as an apprentice or pupil with a firm of mechanical engineers, to recommend an inexpensive and up-to-date book on engineering, one finds some difficulty in making a selection.

To the author it appears that what a beginner really needs is a book which, while giving in broad outlines the information it is necessary to possess concerning the ordinary branches of mechanical engineering, yet shall go sufficiently into detail to enable him to make all the calculations likely to be required during the earlier stages of his career.

It has been the aim of the author in the following pages to state in clear language some of the elementary facts connected with mechanical engineering, and to show how the simple calculations which have to be made from time to time by every engineer and draughtsman can be performed.

Theory is introduced in places where its bearing on practice can be seen and understood. For instance, the theory of raising steam is dealt with after the reader has been introduced to the various types of boiler in use and has learnt something of the merits and demerits of each. Again a somewhat important law of Joule is not stated until the Chapter on Steam Turbines is reached, when the student is able to realise the importance of the law and its application to the De Laval turbine.

The formulæ given are of the simplest character and can be worked out by a beginner having no knowledge of algebra or mathematics, a knowledge of decimals only is assumed.

No attempt is made to describe mechanical processes, such as turning, boring, planing, or iron and brass founding, as the beginner will be able to acquire a knowledge of these processes in his passage through the shops. It is the author's aim to give information which, unless acquired by experience, can only be obtained by reading a considerable number of books dealing with each subject separately.

In actual work many practical questions, such, for instance, as to the speed of a centrifugal pump, or of a high-speed engine, or the heating surface of a given Lancashire boiler, frequently arise. These are usually settled approximately by referring to the makers' catalogues, such catalogues, however, are not often accessible to the beginner, and the author hopes that the information given in this book, much of which he has found useful in his passage through the shops, drawing office, and at the directors' table, may prove of some service to others who are just beginning their career.

R. S. M'LAREN.

January, 1908.

CONTENTS.

CHAPTER I.

Materials.

PAGES

Wrought Iron—Cast Iron—Malleable Iron—Steel—Aluminium—Copper—Gunmetal—Bronzes and Alloys—Testing Materials—Stress—Strain—Elastic Limit—Reduction of Area—Impact Tests—Alternate Bending Tests—Summary of Weights and Strengths—Effect of Temperatures on Strength—Fatigue of Materials—Factors of Safety—Typical Specifications—Cost of Materials, ... 1-18

CHAPTER II.

Bolts and Nuts, Studs, Set Screws.

Bolts and Nuts—Studs—Set Screws—Locking Devices—Lock Nuts—Grover Washer—Helecoid Nuts—Castle Nuts—Steady Pins—Lewis Bolts—Foundation Bolts—Table of Whitworth, and Gas Threads—Rivets—Diameter and Pitch of Rivets for Various Joints, ... 19-26

CHAPTER III.

Boilers.

Boilers—Cornish and Lancashire—Galloway—Economic—Locomotive—Marine—Dry Back—Vertical—Dangers of the Shell Boiler—Grooving, Pitting, and Wasting—Water-tube Boilers—Babcock—Niclausse—Belleville—Stirling—Thornycroft—Yarrow—Strength of Boilers—Steam Raising—Saturated and Superheated Steam—Evaporation—B.T.U. contained in Coal—Rate of Combustion—Draught of Chimneys—Area of Chimneys—Ratio of Grate Area to Heating Surface—Evaporation per foot of Heating Surface—Proper Combustion of Coal—Transmission of Heat through Boiler Plates—Feed-water and Boiler Compositions—Testing Boilers, ... 27-56

CHAPTER IV.

Steam-Raising Accessories.

PAGES

Pumps—Duplex—Weir—Deane—Flywheel Pumps—Injectors—Feed-water Heaters and Economisers—Thermal Storage System — Superheaters — Mechanical Stokers — Bennis—Vicars—Underfeed and Chain Grate—Howden's Forced Draught—Meldrum's Blower—Coal Conveying Plant—Oil Filters, ... 57-70

CHAPTER V.

Steam Pipes and Valves.

Material — Arrangement — Expansion — Size of Pipes — Flow of Steam in Pipes—Radiation from Covered and Uncovered Pipes—Strength of Pipes—Size of Flanges—Water Hammer—Steam Traps—Bucket and Expansion—Exhaust Pipes—Stop Valves — Hopkinson-Ferranti Valve — Isolating Valves, ... 71-88

CHAPTER VI.

The Steam Engine.

Action Explained—Lap—Lead-D-Slide Valve—Piston Valve—Balanced Slide Valve—Reversing Gear—I.H.P.—B.H.P.—Method of Calculating Horse-power—Hyperbolic Curve—Reason why Single Cylinder Engine Uneconomical—Initial Condensation—Compound- and Triple-expansion Engines—Consumption of—Corliss, Willans, Belliss and Browett-Lindley Engines—Merits and Demerits of various Types—Steam Jacketing—Calculating H.P. of Compound Engines—Testing Engines—Brakes—Indicating an Engine and Method of Working out Diagrams, ... 89-127

CHAPTER VII.

The Steam Engine.

Flywheel Calculations—Stored Energy—Stress in Rim—Throttle and Expansion Governors — Proportion of Engine Parts and Stresses — Engine Packings — Hemp, Asbestos, and Metallic—Piston Rings and Springs—Efficiency of Engines—Zeuner Diagram, ... 129-147

CONTENTS.

CHAPTER VIII.

Power Transmission.

Rules for Power Transmitted by Belts—Speed of Belts—Compounding Belts—Pulleys—Balancing and Convexity of Pulleys—Speed of Driving and Driven Shafts—Thickness of Belts for Small Pulleys—Fast and Loose Pulleys—Ropes, Rules for Power Transmitted by—Speed of Ropes—Size of Rope Pulleys—Distance Apart of Pulleys—Clutches for Rope Pulleys—Loss of Power in Transmission—Shafting, Rules for Power Transmitted by—Bearings—Gearing—Toothed Wheels—Helical Wheels—Worm Wheels—Skew Wheels—Raw-hide Pinions—Hans Reynolds' Silent Chain—Rules for Power Transmitted by Gearing, ... 149-163

CHAPTER IX.

Condensing Plant.

Object of Condensing Plant—Jet Condenser—Amount of Cooling Water Required—Air Pumps—Surface Condenser—Extent of Cooling Surface—Amount of Cooling Water Required—Capacity of Air Pump—Temperatures and Corresponding Vacua—Dry and Wet Air Pumps—Vacuum Augmenter—Corrosion in Condenser Tubes—Edwards' Air Pump—Evaporative Condensers—Ejector Condenser—Barometric Condenser—Cooling Towers, ... 165-176

CHAPTER X.

The Steam Turbine.

Parsons Turbine—Willans-Parsons Turbine—Brush-Parsons Turbine—Speeds and Outputs—De Laval Turbine—Curtis Turbine—Rateau Turbine—Westinghouse Turbine—Zoelly Turbine—General Remarks, ... 177-192

CHAPTER XI.

Electrical Chapter.

Production of Current — Magnetic Field — E. M. F. — Simple Dynamo—Volts, Amperes, Ohms, and Watts—Electrical Horse-power—Power and Current Required for Lamps—Reason for High Voltages—Alternators—Transformers—Single and Multiphase Currents—Series-wound, Compound-wound, and Shunt-wound Dynamos—Rotary Converters—Motor Generators—Primary Batteries and Accumulators—Summary of Electrical Terms—Table of Conductors, . 193-209

x   CONTENTS.

CHAPTER XII.
Hydraulic Machinery.

PAGES

Hydraulic Press and Hand Pump—Hydraulic Accumulator—
Pressures Used by Hydraulic Engineers — Hydraulic
Riveters, Fixed and Portable—Flanging Press—Lifts and
Cranes – Jiggers—Jack—Water Wheels—Overshot, Undershot, and Breast Wheels—Turbines—Parallel Flow—Radial
Flow—Mixed Flow—Advantages and Disadvantages of
each—Impulse Wheels—Power obtained from Falling
Water—Suction Tube—Governing—Bearings—Niagara
Falls Turbines—Falls of Foyers Turbines—Pumps - Centrifugal—Multiple-Stage Pumps—Sizes and Speeds—
Reciprocating Pumps—Pumping Hot Water—Valves—
Pulsometer Pump—Hydraulic Ram—Materials for Pumping Various Liquids—Flow of Water in Pipes and Useful
Memoranda, ...   211-237

CHAPTER XIII.
Gas and Oil Engines.

Otto Cycle – Description of 10-H.P. Engine – Methods of Ignition
—Exhausting and Scavenging—Drawing in Mixture—Compression—Exceptions to Otto Cycle—Körting Engine—
Amount of Gas Consumed and B.T.U. Contained in Gas—
Horse-power of Gas Engines—Temperatures and Pressures
Reached—Speeds—Thermal Efficiency—Distribution of
Heat—Suction Gas Plant—Amount of Gas Made and
Water Required—Starting Gas Engines—Comparative
Merits of Steam and Gas Engines—Oil Engines—Consumption of Oil and Petrol, ...   239-254

CHAPTER XIV.
Strength of Beams and Useful Information.

Strength of Beams having Top and Bottom Flanges—Strength of
Rectangular Beams—Table of Rolled-steel Joists—Ascertaining Stresses by Graphic Methods—Calculating Power
Transmitted by Screws, Levers, and Wedges—Thermometer Scales—To Divide a Straight Line into a Number
of Equal Parts—Decimal Equivalents of an Inch with
Areas and Circumferences—Table of Areas and Circumferences—Table of Squares and Cubes—Weights and
Measures with Metrical Equivalents, ...   255-269

CHAPTER XV.

Conclusion, ...   271-276

Index, ...   277

LIST OF ILLUSTRATIONS.

FIGS.		PAGE
1.	Incorrect method of ribbing a flat plate,	11
2.	Correct ,, ,, ,,	11
3.	Bolt and nut,	19
4.	Stud,	19
5.	Set screw,	19
6.	Lock nuts,	20
7.	Grover washer,	20
8.	Castle nut,	20
9.	Helicoid nut,	20
10.	Set screw and steady pin,	20
11.	Lewis bolt,	20
12.	Foundation bolt and anchor plate,	21
13, 14.	Rivets, before and after closing,	24
15.	Lancashire boiler,	28
16.	,, front view with setting,	29
17.	Cornish boiler, front view,	32
18.	Locomotive boiler,	34
19.	Marine ,,	35
19a.	"Blake" ,,	36
20.	Babcock ,,	38
21.	Worthington Duplex pump,	57
21a.	Weir feed pump,	59
22.	Cameron feed pump,	60
23.	Diagram showing action of injector,	62
24.	Modern injector,	63
25.	Feed-water heater,	64
26.	Underfeed stoker,	68
27.	Dangerous pipe arrangement,	72
28.	Correct position of stop valve on boiler,	73
29.	Incorrect ,, ,, ,,	73
30.	Stop-valve passage obstructed by water,	77
31.	Bucket steam trap,	83
32.	Expansion ,,	83
33.	Screw-down stop valve,	85
34.	Straight-through gate stop valve,	86
35.	Vertical single-cylinder steam engine,	90
36.	Cylinder with piston valve,	92
37.	Reversing gear,	92
37a.	Setting eccentrics,	92
38.	L.P. cylinder with balanced slide valve,	93
39, 40, 41.	Hyperbolic curves,	99
42.	Corliss gear,	105
43.	L.P. cylinder with Corliss valves,	105
44.	Drop-valve engine,	107
44a.	Van den Kerchove engine,	107
45.	Willans triple-expansion engine,	109

LIST OF ILLUSTRATIONS.

FIGS.		PAGE
46.	Crosby indicator,	121
46a.	,, ,, for superheated steam,	121
47.	H.P. Indicator diagram,	124
48.	L.P. ,, ,,	124
49.	Metallic packing,	141
50, 51, 52.	Ramsbottom piston rings,	143
53, 54, 55.	Willans piston rings,	143
56, 57.	Mudd piston rings,	143
58.	Zeuner diagram,	146
59.	,, with piston and cylinder,	146
60, 61.	Worm and wheel,	160
62.	Hans Reynolds' chain,	162
63.	Jet condenser,	166
64.	Air pump,	167
65.	Surface condenser,	168
66.	Condenser tube end and ferrule,	169
67, 68.	Edwards' air pump,	172
69.	Willans-Parsons steam turbine,	178
70.	Turbine blading,	179
71.	,, baffle rings,	179
72.	Method of strengthening Parson's blades,	179
73.	Bucket wheel of De Laval turbine,	183
74.	De Laval turbine with gearing,	184
75.	Curtis turbine blading,	186
76.	,,	187
77.	Horse-shoe magnet,	194
77a.	,, ,,	194
78.	Direct current 12-pole dynamo,	199
79.	Alternator,	201
80, 81, 82, 83.	Alternating-current diagrams,	202
84.	,, ,, ,,	203
85.	,, ,, generator,	204
86.	Hydraulic press,	212
87.	Portable hydraulic riveter,	212
88.	Hydraulic U leather,	212
89.	,, accumulator,	213
90.	,, riveter,	215
91.	,, flanging press,	218
92.	,, punching machine,	218
93.	,, jigger,	219
94.	,, jack,	219
95.	Water turbine, Jonval type,	222
96.	,, radial inward-flow type,	223
97.	,, ,, outward-flow type,	224
98.	Centrifugal pump,	229
99.	Pulsometer ,,	233
100.	Gas engine,	240
101.	Korting gas engine,	243
102.	Suction gas producer,	248
103.	Hornsby oil engine,	253
104, 105, 106.	Graphic methods of computing stresses,	260

MECHANICAL ENGINEERING
FOR BEGINNERS.

CHAPTER I.

MATERIALS.

The materials chiefly used in mechanical engineering are—wrought iron, cast iron, steel, aluminium, gunmetal, and kindred alloys. Iron may, for practical purposes, be divided into two distinct classes—viz., wrought and cast—and although the difference between them lies principally in the fact that there is less carbon and silicon in wrought iron than in cast iron, yet in practice the distinction between the two forms of iron is so marked that they might almost be two distinct metals.

Wrought Iron is fibrous, can be bent (or given a permanent set) without breaking; it can be sheared and punched; when hot it can be worked under the hammer, and forged into various shapes. Two pieces of this metal can be joined or welded together by heating them to what is known as welding heat (about 1,600° F.) and hammering the two parts together. The surface of wrought iron can be made exceedingly hard, if required, by case-hardening. This is effected by surrounding the metal with cuttings of leather, horn, and bone dust, raising the whole to a considerable heat, maintaining the heat for some hours, and then slowly cooling. The carbon from these chippings enters into the composition of the metal, and forms a surface closely allied to steel. Case-hardening can also be effected by heating the metal to a bright red and applying prussiate of potash, allowing to cool slightly, and then plunging into cold water. This method is quicker than the one previously mentioned, but is more liable to cause distortion.

Wrought iron is used in cases where considerable strength is necessary, and the required form is a simple one, such as can be obtained by forging, welding, and machining the metal, stamping out of the solid, or where the structure can be built up of plates, Angle and Tee irons. During recent years wrought iron has been almost entirely superseded in mechanical engineers' works by mild steel, but the former is still used in cases where it requires to be welded, as in chains, tubes, eye-bolts, &c. In cases where shocks have to be withstood, wrought iron is superior to mild steel, as the former has not the same tendency to crystallise or lose its fibrous character. The coupling links of railway waggons are always made of the best wrought iron.

Wrought iron and mild steel are brought into an engineer's works in the following forms, viz.:—Round, square, and flat bars; sheets or plates, Tee, Angle, and Channel iron. The metal is formed into these shapes by being passed through rolls when hot.

Wrought iron varies considerably in quality; its distinguishing marks, in addition to the maker's brand, are—B, BB, and BBB; the letter B signifying the word "best."

Wrought iron of good quality will safely bear a stress of 5 tons per square inch in simple tension—that is to say, a bar 1 inch square will safely support any weight, up to 5 tons, suspended from it. If, however, the load is likely to be a varying one, say from 0 to the maximum, then the bar should not be loaded to more than $3\frac{1}{2}$ tons per square inch. The breaking stress of the best Yorkshire bars in simple tension is about 28 tons per square inch, and of Staffordshire bars about 25 tons per square inch. The elongation before fracture varies from 8 to 30 per cent. on a test piece 10 inches long. Wrought iron will safely bear a stress of 4 tons in compression—*i.e.*, a block 1 inch square will safely support a weight of 4 tons placed upon it. Wrought iron will begin to crush, or give, under a pressure of 13 to 20 tons per square inch. If subjected alternately to tension and compression, wrought iron should not be loaded to more than 2 tons per square inch.

A cubic inch of wrought iron weighs approximately ·28 lb. The weight of any piece or structure of wrought iron is, therefore, easily ascertained by finding the number of cubic inches it contains and multiplying by ·28.

Wrought iron melts at a temperature of from 2,700° to 2,920° F.* It contains not more than ·2 per cent. of carbon.

Cast Iron is granular and of a brittle nature; it cannot be bent, sheared, or punched, and is liable to break suddenly if too

---

* The melting point of metals is not easy to determine with great accuracy. Mercury boils at 680° F., and, therefore, cannot be used. Very high temperatures are usually determined by noting the resistance caused to an electric current when passed through a platinum wire, the latter being placed in a porcelain or quartz casing. The higher the temperature the greater the resistance.

great stress is put upon it. Cast iron, however, can be run, when in a molten state, into intricate shapes. Two pieces of this metal, after working together for some time in sliding contact, acquire very smooth surfaces, and the amount of friction between them is less than with any other two similar pieces of metal.

Cast iron is used in all cases where an intricate form is required and sufficient metal can conveniently be given to withstand the stresses likely to come upon it, and where corrosion can be prevented by the friction of the parts or by other methods. Cast iron is used for steam cylinders, bedplates, and frames of various machines, for flywheels, large water pipes, railway chairs, and for an enormous variety of purposes. Cast iron is brought into an engineer's foundry in the form of pig iron: when required for use it is broken into smaller pieces and melted down, together with a certain proportion of old broken cast iron (scrap), when this can be obtained, in a cupola.

The cupola consists of an upright wrought-iron casing lined with fire-brick, fireclay, and a fire-resisting substance called Ganister. The fire having been kindled, the cupola is charged with coke and pig iron (broken) in alternate layers, and a blast of air from a fan or blower is driven in, the iron melts and trickles down through the coke to the bottom of the cupola. It is then drawn off in a liquid state and poured into moulds made to any desired shape in sand. While the first charge is being drawn off and used, a further supply of coke and iron is thrown in at an opening at the top of the cupola.

Pig iron is obtained in differing degrees of hardness and strength; these degrees are known by numbers. No. 1 is very soft and easy to machine, but is deficient in strength, and is to a certain extent porous. No. 2 is slightly harder than No. 1. No. 3 pig is hard, tough, and close grained; it is largely used for good castings. Nos. 4, 5, and 6 are still harder and tougher than the foregoing. The degree of hardness of old cast-iron scrap depends upon the quality of the original metal and the number of times it has been melted down. Scrap is cheaper than pig iron, but castings, if made entirely from this source, would be deficient in strength, very brittle, and extremely hard to machine.

The cast-iron parts of machines that are subject to wear, such as steam cylinders (in which the piston moves at a considerable speed), should be made much harder than is necessary for those parts which are not subject to wear, such as bedplates, &c. For steam cylinders a certain proportion of No. 4, and even No. 5 pig, is added to a mixture of No. 3 pig and scrap iron. For castings of an inferior kind it is customary to use a large proportion of scrap and a very small proportion of No. 1 pig iron. Every time, within certain limits, cast iron is remelted it becomes harder and stronger, but more brittle. Small castings can be made exceedingly hard by pouring the molten metal into iron moulds or chills instead of into sand moulds. Castings treated in this way are called "chilled."

Cast iron, if of really good metal, thoroughly sound, and free from internal stresses set up by unequal contraction in cooling, will safely bear a stress of $1\frac{1}{4}$ to $1\frac{1}{2}$ tons per square inch in simple tension, but in cases where the stress varies with considerable frequency from 0 to maximum, the metal should not be loaded to more than 1 or $1\frac{1}{4}$ tons per square inch. In cases where the metal is subjected to alternate tension and compression, the working stress should not exceed $\frac{1}{2}$ a ton per square inch.

The ultimate or breaking tensile stress of cast iron varies from 8 to 14 tons. It will safely bear a stress of 6 tons in compression. The ultimate crushing stress is from 25 to 50 tons per square inch. A cubic inch weighs from ·26 to ·27 lb. The melting point of cast iron is about 2,050° F. in the case of No. 1 pig. Nos. 4 and 5 pig require a temperature of about 2,250° before melting. Cast iron contains from 3 to $3\frac{1}{2}$ per cent. of carbon and from $1\frac{1}{2}$ to $2\frac{1}{2}$ per cent. of silicon. Not more than $1\frac{1}{2}$ per cent. of the carbon is combined with the iron, the remainder is present in the form of graphite.

The transverse strength of cast iron is usually ascertained by placing a bar, 2 inches deep by 1 inch wide, upon supports 36 inches apart, and loading it in the centre. A bar tested in this way will break with a load of from 25 to 40 cwts., depending upon the quality of the iron. In such a test the molecules in the upper part of the bar are in compression, while those in the lower part are in tension. A typical specification is given at the end of the chapter.

Malleable Cast Iron is not so brittle as ordinary cast iron; it will bend slightly before breaking. Ordinary castings can be made slightly malleable by surrounding them with a substance which will extract some of the carbon, such as crushed red hematite, or black oxide of iron (iron scale), and placing them in an oven maintained at a temperature of 1,800° to 2,000° F. during the day and at a dull red heat during the night for several days and nights, the length of time depending upon the size of the casting. If the castings are required to be very malleable, they are made of the best Cumberland white or grey pig iron. Such castings, before being treated, are extremely brittle. Malleable castings are used in cases where shocks may have to be withstood, and where the form renders it difficult or expensive to make the article of wrought iron, and where the expense of gunmetal or bronze has to be avoided.

Steel.—In steel there is not the same hard and fast division between cast and wrought as there is between cast and wrought iron, but for the purpose of classification finished steel may be divided into three classes—viz., mild, cast, and tool steel.

Mild steel has many of the properties of wrought iron, but is stronger. It can be bent, sheared, forged, and, if it has a very small percentage of carbon, can be welded. Holes in steel plates, unless the plates are very thin, should be drilled and not punched, as punching injures the surrounding metal. Mild steel can be cut or machined when cold by tool steel. Mild steel usually enters an engineer's works in the form of round and square bars, plates, and rough forgings. It is used in cases where greater strength and hardness are required than are obtained with wrought iron, but where in other respects the latter would be used.

Mild steel is largely used for crank shafts, piston-rods, eccentric-rods, boiler plates, cross-head pins, studs, rivets, &c. Steel containing a small proportion of nickel is stronger than ordinary mild steel, and does not rust so easily, but is more difficult to machine. Mild steel will safely bear a stress of about 6 tons in simple tension or in compression, or 4 tons when the load varies from 0 to maximum, but when subject alternately to tension and compression, as in a piston-rod of a double-acting engine, $2\frac{1}{2}$ tons per square inch is a sufficient load if a long life is desired. The breaking tensile stress of mild steel is about 30 tons, and the elongation 20 to 35 per cent. on 10 inches. The torsional strength of mild steel is dealt with in the chapter on transmission of power, and some remarks as to its ability to withstand shock will be found later on. The weight of a cubic inch is ·288 lb. Mild steel contains from ·1 to ·5 per cent. of carbon. It does not appreciably harden if heated and plunged in water.

Cast Steel is harder and stronger than cast iron, and will bend before breaking; in fact, the best qualities are improved by being forged. Great care has to be exercised in making steel castings, as the metal in cooling gives off a gas, which, unless got rid of, honeycombs the casting with small holes. Molten steel is not so fluid as iron, and does not fill the cavities so completely as the latter. Messrs. Whitworth cast some of their steel under pressure in order to overcome these difficulties.

Cast steel is used in cases where great strength is required, and where the form is only moderately intricate. It is used for toothed wheels, hydraulic cylinders, high-pressure valve bodies, for the arms of hydraulic riveters, &c. Steel castings, if sound, will safely bear a stress of from 4 to 6 tons per square inch in simple tension or compression. The breaking tensile stress is from 20 to 30 tons, with an elongation of 15 to 25 per cent. on 2 inches. Whitworth fluid-compressed steel has an ultimate breaking strength of about 40 tons per square inch, with an elongation of about 32 per cent. The weight of a cubic inch is ·288 lb. Steel castings usually contain about ·251 per cent. of carbon. The melting point is from 2,400° to 2,600° F.

Tool steel is much harder than mild steel, and is more difficult to forge. Tool steel usually enters an engineer's works in square, hexagonal, or round bars; it is cut and forged to the required shape and size, and is then hardened and tempered to make it sufficiently hard to cut mild steel, cast iron, or wrought iron.

Hardening is effected by heating the piece of steel to a cherry red (1,600° to 1,800° F.) and then plunging it into cold water or oil. A piece of steel treated in this way only would be too hard and brittle for many purposes. It is tempered by being again heated to a temperature of between 430° and 570° F. and then suddenly plunged into cold water. If the piece of steel is of adequate size the hardening and tempering can be done at one heating. The tool is heated to a cherry red, the part to be hardened only is plunged into water and cooled; sufficient heat then travels along the steel from the uncooled portion to raise the temperature of the cooled part to the required degree, when the whole piece is plunged in water. The temperature is known by the colour of a film of oxide which appears on the bright steel.

Self-hardening Tool Steel.—If a tool, after being hardened and tempered, is reheated and allowed to cool slowly it loses its temper and becomes soft. When a tool is used for cutting hard materials at a high speed the tool becomes hot, and if the heat reaches a certain point the tool loses its temper. It was found, however, that by adding tungsten and a small percentage of manganese and chromium the steel became self-hardening— that is to say, it did not become soft when heated and allowed to cool. Mushet steel was the best known example of this class of steel.

High-speed Tool Steel.—A still further advance in steel used for cutting purposes has been made in what is known as high-speed tool steel. In it the percentages of tungsten and chromium are increased, and that of the carbon diminished. The steel is heated almost to melting point, then cooled by an air blast; then reheated to a dull red and again cooled by an air blast. In hardening and tempering these steels in the works, the maker's directions, which are sent out with every bar, must be followed. Quenching in water is forbidden.

The percentage of carbon in tool steel varies considerably. In ordinary cutting tool steel there is about 1 per cent.; in Mushet steel about 2·15 per cent.; in high speed tool steel about from 1·25 down to ·6 per cent. The ultimate tensile strength of tool steel is from 40 to 60 tons per square inch. The elongation is 5 to 12 per cent. on 2 inches.

As steel is often referred to as being of the Bessemer or open hearth process, it may be well to state briefly what these processes are. Steel is obtained from pig iron by removing a large proportion of its carbon, and as much of the silicon, phosphorus, and sulphur as possible. In the Bessemer process this is effected by blowing air through the molten pig placed in a converter; as a result the silicon is first burnt out and forms slag, which is removed; the carbon then burns, making a long flame of carbon monoxide; when all the carbon is burnt the flame ceases and a certain amount of Ferro-manganese or Spiegeleisen is added to give the required percentage of carbon to the steel.

In the Siemens-Martin open hearth process, pig iron and wrought-iron scrap are melted together and a proportion of hematite added; the air is not blown through the metal as in the Bessemer process, but a mixture of air and producer gas are burnt above the molten metal. As the pig iron is mixed with scrap iron the amount of silicon and carbon is not so great as in a corresponding quantity of pig iron alone, and what silicon and carbon there is is got rid of by oxidation caused by the burning gases, aided by the oxygen of the hematite. If the whole of the carbon is removed a certain amount of ferro-manganese is added as in the Bessemer process.

The Siemens open hearth process is similar to the Siemens-Martin, but in the former pure pig iron and not pig and scrap is used.

Briefly, it may be stated that the Bessemer process is chiefly used for making high carbon steel, and the open hearth process for mild steel.

One frequently reads of "acid" and "basic" steel; these terms mean that the lining of the open hearth furnace or converter was made either of an acid or basic material. The lining of the furnace has a certain effect in removing impurities from the steel.

Steel which contains phosphorus is brittle and is thought to deteriorate in time, especially if subjected to stresses of an alternating kind.

A small amount of nickel, as already stated, increases the tenacity of steel. A small percentage, too, of vanadium has a beneficial effect; it appears to cause the carbon to distribute itself more evenly, and adds to the ability of the steel to resist shock.

Aluminium is a metal which is used in cases where lightness is required; it can be cast into intricate shapes. It is very soft, but has a fair tensile strength, approximately that of cast iron. Its ultimate breaking strength in simple tension is about 8 tons per square inch, and may in practice be loaded to 1 ton per square inch. An alloy formed of aluminium and zinc is still stronger. Aluminium bronze is dealt with under *alloys*. A cubic inch weighs ·09 lb. The melting point is about 1,200° F.

Copper is fibrous and very soft; it can be bent, sheared, or hammered into various shapes when cold. After hammering it requires to be annealed—*i.e.*, raised to a high temperature and then suddenly cooled.* Copper is a good conductor of heat and does not readily corrode. Two pieces of copper can be joined together by brazing: brazing is effected thus—the two pieces to be united are placed together, brazing metal in the form of borings is placed over the joint and the whole is covered with powdered borax; the seam is then held over a coke fire or gas flame until the brazing metal melts and unites the two pieces of copper. Borax, which acts as a flux, is thrown on during the operation. A joint which has been properly brazed is as strong as the metal itself.

Copper, on account of its heat-conducting properties, is used in the construction of locomotive fire boxes. It is also an extremely good conductor of electricity and is largely used by electrical engineers. Copper is also used for small pipes which require to be bent cold.

Copper will safely bear a tensile stress of $1\frac{1}{2}$ tons per square inch at temperatures below 300° F.; above this temperature the strength rapidly falls off. The breaking tensile stress of copper bolts and plates at 300° F. is about 14 tons per square inch, with an elongation of about 35 per cent. on 2 inches. At 600° F. the strength falls to about 10 tons per square inch. The melting point of copper is about 1,950° F. If the heat-conductivity of gold is taken as 100, that of copper is about 87, while iron or steel has a heat-conductivity of about 37 only. A cubic inch weighs ·322 lb.

Gunmetal, Bronzes, and Kindred Alloys.—Gunmetal is

---

* When steel is annealed it is raised to a high temperature and *slowly* cooled.

stronger than cast iron and will bend slightly before breaking. It is soft and easy to machine, can be cast into intricate shapes, is not porous, and does not readily corrode, but is expensive; it cannot be forged or welded. Gunmetal is used in cases where comparatively small castings are required to be of an intricate form and of moderate strength, also in cases where corrosion has to be avoided. It is used largely for the internal parts of steam and water valves, for cocks, and engine fittings.

Gunmetal is largely used for bearings in which steel shafts revolve, as the friction between this alloy and steel is not excessive, and gunmetal, being the softer, wears away before the steel: when worn to too great an extent the "brasses" can be replaced. Bearings that are subject to heavy and continuous loads are usually lined with white metal or anti-friction metal.

Gunmetal is composed of copper, tin, and zinc; the amount of each is varied slightly according to the purpose for which the alloy is required. The proportions adopted by the Admiralty are—copper, 88 parts; tin, 10 parts; zinc, 2 parts. By adding more tin the alloy is made harder. The component metals, when the required casting is of small or medium size, are melted down in plumbago pots or crucibles; when a large casting is required the metals are melted down in a furnace.

Gunmetal will safely bear a simple tensile stress of $1\frac{1}{2}$ to 2 tons per square inch. The breaking stress is from 10 to 16 tons, with an elongation of $7\frac{1}{2}$ to 10 per cent. The weight of a cubic inch is ·3 lb. The melting point is about 1,800° F.

Phosphor Bronze has all the properties of gunmetal, but is stronger; when hot it can be forged and rolled into rods. It is composed of copper, tin, and phosphorus, the Admiralty proportions being copper 83, tin 10, and phosphide of copper 7. Phosphor bronze is used for valve spindles, pump-rods (where steel should not be used on account of corrosion), bearings, and for small castings in cases where greater strength is required than can be obtained by the use of gunmetal.

Phosphor-bronze rods and forgings will safely bear a tensile stress of 4 to 5 tons per square inch, and castings from 2 to 3 tons. The breaking stress is from 10 to 30 tons, with an elongation of 10 to 30 per cent. on 2 inches. The weight of a cubic inch is ·3 lb.

Manganese Bronze, Stone's Bronze, and **Delta Metal** have all the advantages of and are stronger than phosphor bronze; in fact, they can be made stronger than mild steel, but at the expense of ductility. All these bronzes contain a small amount of manganese, iron, and other hardening materials; the exact proportions are, however, a trade secret. These bronzes are largely used for pump-rods, valve spindles, ships' propellers, and in cases where a strong non-corrodible rolled rod or casting is required. Manganese bronze, Stone's bronze, and Delta metal rods will safely bear a tensile stress of 5 to 7 tons per square inch, and castings from 3 to 7 tons. The breaking tensile stress of these rods varies from 29 to 35 tons, with an elongation of 10 to 32 per cent. on 2 inches.

Aluminium Bronze.—An exceedingly strong and ductile bronze is made by a mixture of 90 parts of copper and 10 parts of aluminium. The ultimate breaking stress of this alloy, when cast, is about 30 tons per square inch in simple tension, with an elongation of 20 per cent. on 2 inches. When rolled, a tensile strength of 38 tons, with an elongation of 28·8 per cent. on 2 inches, has been obtained. If, however, the alloy is subjected to a temperature of about 570° F., the strength falls off and the alloy becomes brittle. At a temperature of 750° F. the alloy is so brittle as to be of little use. The weight of a cubic inch is ·273 lb.

Muntz Metal is more ductile, but not quite so strong as the bronzes just mentioned. This metal consists of from 60 to 62 per cent. of copper and from 40 to 38 per cent. of zinc. It is very suitable for condenser tubes, and for bolts which have to resist the corrosive action of sea water. The safe tensile stress for Muntz metal rods is about $5\frac{1}{4}$ tons per square inch, the breaking stress about 26 tons, with an elongation of 45 per cent. on 2 inches.

Brass is inferior in strength to, and more brittle than, the above-mentioned alloys. It is composed of copper and zinc, with a small proportion of tin. The safe tensile stress is about 1 ton, or less, per square inch.

White Metal, sometimes called Babbit metal, from the name of the metallurgist who first introduced it, is very soft, and has valuable anti-friction properties—*i.e.*, it remains cool when subject to rubbing contact under heavy pressure, and for this reason it is largely employed for lining bearings. White metal is composed principally of tin, with an addition of antimony and copper. The Admiralty proportions are tin 85 to 89, copper 7 to 2, and antimony 8 to 9. A cheaper form of white metal is made of lead and a small proportion of antimony, but is not considered to be so good. White metal is usually bought in ingots from firms who have made a study of the subject. The weight of anti-friction metals made from tin and antimony is about ·26 lb. per cubic inch. The weight, when made of lead and antimony, is about ·4 lb. per cubic inch. With regard to the strength, Professor Goodman has recently found that the ultimate strength of some white metal, consisting of lead 90 parts and antimony 10 parts, was about $3\frac{1}{4}$ tons in tension and $7\frac{1}{2}$ tons in compression.

In making castings of iron, steel, gunmetal, or of any other alloy, it is essential for strength that sharp corners should be avoided, for in cooling the crystals set in a direction at right angles to the face of the casting, and every sharp corner constitutes a source of danger. Any sudden increase, too, in the thickness of the metal should be avoided. It has been found, for example, that a flat plate strengthened by ribs running into a common centre or small mass of metal, as shown by Fig. 1, is

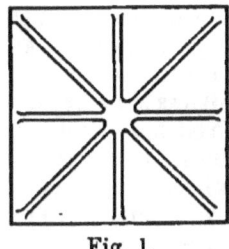 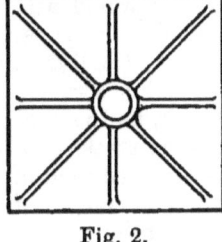

Fig. 1.   Fig. 2.

Incorrect and correct methods of ribbing a flat plate.

not so strong as when strengthened in the manner shown by Fig. 2, where the cooling of the metal can take place more evenly. A good radius should be given to the ribs as they meet the central ring. Pulleys and small flywheels, in which the section of the rim differs greatly from that of the arms, usually have the latter cast in the form of an S. This form allows a certain amount of give while the metal is cooling.

Testing Materials.—Until comparatively recent years it was considered that if the tensile strength and percentage of elongation of a certain metal, or alloy, was known, a judgment could be formed as to whether such metal or alloy was, or was not, suitable for a given purpose. It has, however, been found that of two pieces of steel giving equally good results both as to strength and elongation, one may be capable of withstanding shock very much better than the other, also that one may be able to withstand reversal of stress better than the other. Tests to ascertain the behaviour of materials under shock and reversal of stress are now frequently made, but before describing the methods of making these tests it may be well to state what takes place when a piece of metal is tested to destruction in an ordinary testing machine, so that the beginner may have clear ideas as to terms, such as stress, strain, elastic limit, yield point, reduction of area, &c. When a small rod of metal, or test piece, is put into a testing machine and its ends pulled apart, the force applied produces "stress" in the metal and the metal stretches; the amount by which the metal stretches, or its deformation, is called "strain." Up to a certain point metal retains its elasticity, so that if the load is removed the metal returns to its original form. When the stress has reached a certain point the strain or deformation increases out of proportion to the stress, and the metal, provided it does not fracture like cast iron, takes a permanent set; the point at which this occurs is called the elastic limit or yield point.* Naturally the safe working stress of the metal or alloy is well below the elastic limit. If still further force is applied after the metal has reached its elastic limit, the rod becomes permanently stretched, and its area becomes reduced; finally, when a certain stress is reached the rod fractures. This stress is the ultimate or breaking stress of the material, but, as the area of the rod has become reduced, the ultimate breaking stress can be stated in two ways—either as so many tons per square inch of original section, or as so many tons per square inch at point of fracture. The former is the most useful information for practical engineers, the latter for scientific investigators into the properties of materials.

The amount by which the rod has stretched is called its "percentage of elongation." Thus, if the working length of the test piece is 2 inches, and it stretches $\frac{1}{2}$ inch before fracture, its elongation is 25 per cent. As the stretching is greater at the point of fracture than in the other parts of the rod, the percentage of elongation on a long test piece is less than on a short one, hence the length of the test piece should be given at the same time as the percentage of elongation if the information is to be of service.

Testing machines have been designed and are in use in many Technical Colleges and Laboratories, which, by means of multiplying levers and pencil, draw automatically a diagram showing the actual deformation that takes place in the metal when subjected to gradually increasing stress. Such diagrams are called stress-strain diagrams.

Impact Tests.—In order to ascertain whether a given metal or alloy is suitable for resisting shock, an impact test should be made. A very convenient and simple machine for this purpose was designed a few years ago by Mr. Izod and is made by Messrs. Avery. Izod's impact testing machine consists of a hammer suspended at the end of a swinging rod. The sample piece of

---

* The true yield point occurs slightly later than the point of elastic limit, but for ordinary purposes they may be looked upon as identical.

steel or alloy to be tested is of small rectangular section and is nicked to a uniform depth at the place where fracture is desired; the test piece is placed in a vice immediately under the hammer, the latter is then swung back to a given distance and released; in falling the hammer strikes the test piece and fractures it. The distance traversed by the hammer (shown by a dial and pointer), after fracturing the test piece, shows the resistance of the latter. When the steel or alloy under test is brittle the hammer travels much farther after fracture than when the material is suitable for withstanding shocks. In this connection it may be mentioned that if mild steel is heated up to about 1,330° F., and then quenched in oil, its ability to withstand shock is enormously increased.

A simple testing machine for ascertaining the ability or otherwise of metals to withstand alternate bending has recently been devised by Captain Sankey. The machine, which is made by Messrs. Casella, consists of a fixed and a moving vice, the latter being provided with a long handle. The piece of metal to be tested is gripped by the fixed vice and bent by means of the moving vice and handle. The moving vice is arranged so that the pressure exerted in bending the test piece is transmitted through springs, and by an arrangement of ratchet, dial, and pencil, a line or rather arc, indicating the pressure exerted, is drawn every time the metal is bent, so that both the number of times the metal bends before fracture and the pressure exerted each time are automatically recorded. The results of tests made with this machine agree very closely with tests made with a much more elaborate machine constructed by Professor Arnold.

Table I. gives a rough summary of the safe and ultimate tensile, compression, and shearing stresses of various materials, with the percentage of elongation, also the weight of a cubic inch.

Some tensile and compression tests, made by Mr. Izod, upon four pieces of crucible steel are embodied in the table, as they show clearly how the tensile strength of steel increases with an increased percentage of carbon, while the elongation falls off. With regard to the bronzes enumerated, the higher tensile strengths are only obtained at the expense of ductility—i.e., as the strength increases the percentage of elongation falls off.

It may be noticed that while the ultimate shearing stress of wrought iron and steel is less than the ultimate tensile stress; in the case of cast iron the ultimate shearing stress is greater than the ultimate tensile stress. The figures, which are based upon trials recently made, are confirmed by some tests[*] made by

---

[*] Communicated to the Institution of Mechanical Engineers, in connection with a paper on "Shear," by Mr. E. G. Izod, 1905.

TABLE I.

	Safe Stress.	Simple Tension. Breaking Stress.	Elongation on 2 Inches.	Elastic Limit.	Simple Compression. Safe Stress.	Breaking Stress.	Shearing. Breaking Stress.	Weight of a Cubic Inch.
	Tons per Sq. Inch.	Tons per Sq. Inch Original Section.	Per Cent.	Tons per Sq. Inch Original Section.	Tons per Sq. Inch.	Tons per Sq. Inch Original Section.	Tons per Sq. Inch Original Section.	
Wrought iron,	5	18-26	8-50	10-15	4	13-20	13-21	·28
Cast iron,	1¼	8-14	...	8-14	6	25-50	10-18	·26
Mild steel,	6	26-32	20-50	14-18	6	25-30	20-25	·28
Swedish crucible steel—								
0·12 carbon,	...	24·9	43·0	...	...	...	18·5	·28
0·48 ,,	...	42·1	26·0	...	...	...	28·8	·28
0·71 ,,	...	56·3	15·0	...	...	...	36·6	·28
0·77 ,,	...	61·3	11·0	...	...	...	38·3	·28
Steel castings,	5	20-30	15-25	11-16	6-8	30-40	25-30	·28
Aluminium,	1	6-8	25	6	1	4-8	3-5	·09
Do. alloy,	2	12	10	...	...	...	7	...
Copper—Sheets and rods,	1¼	14-20	20-45	3½-4	1¼	8-10	6-9	·322
Gunmetal,	1½	10-16	7½-10	5-8	1½	8-15	10-16	·3
Phosphor bronze,								
Manganese ,, Rolled,	4-7	25-35	10-32	13-25	5	25-30	22-30	·31
Aluminium ,, Cast,	2-7	15-35	10-20	10-20	6	30-60	16-25	·298
Stone's ,,								
Delta metal,	5¼	26	45	11	4	18-20	15	·3
Muntz metal,	1	5-8	5-7	4	1	6-8	5-8	·29
Brass,	...	...	...	...	...	...	...	·26
White metal (see p. 11),	...	...	...	...	...	...	...	·41
Lead,	...	...	...	...	...	...	...	

Professor Goodman of Leeds, and given in Table II. This table is instructive, as showing the reduction in area which takes place when certain metals and alloys are tested to destruction, and how different some of the results appear when they are given in terms of "tons per square inch of original section," and when given in "tons per square inch at point of fracture."

TABLE II.—TESTS MADE BY PROFESSOR GOODMAN AT LEEDS.

	Ultimate Tensile Stress— Original Section.	Ultimate Tensile Stress at Point of Fracture.	Reduction in Area.	Ultimate Shearing Stress— Original Section.
	Tons per sq. in.	Tons per sq. in.	Percentage.	Tons per sq. in.
Cast iron,	10·9	10·9	Nil.	12·9
Gunmetal (soft and ductile),	14·9	19·5	30·7	14·2
,, (hard and brittle),	12·4	12·8	3·0	17·4
Moderately hard steel,	48·0	49·8	7·4	34·0
Mild steel,	23·6	29·5	67·5	18·9
Wrought iron (soft),	21·7	27·4	49·8	17·4
,, (merchant),	22·6	24·7	24·6	24·5
Copper (annealed),	14·8	23·7	65·0	11·0
,, (hard-drawn),	17·2	24·1	47·0	23·3
Aluminium,	8·8	12·7	45·5	5·6

The tensile strength of steel and of most of the bronze alloys can be increased by cold rolling—*i.e.*, reducing the section of the rod by passing it through rolls. The increased tenacity is, however, gained at the expense of ductility. Copper and iron wire, when drawn down by successive operations, is much stronger than when in the form of rod. Copper wire will bear a stress of about 25 tons per square inch (as against about 14 tons in the case of copper rods or plates) even after annealing; before annealing the wire has even greater tenacity.

Fatigue of Metals.—It has been found from test pieces cut from material which has been subjected for some time to reversal of stresses and to shock, that its percentage of elongation has fallen off from the original figure. This falling off in ductility and real strength is sometimes referred to as the "fatigue of the material." Wöhler found out over thirty years ago that frequent reversal of stress—*i.e.*, from tension to compression, or *vice versâ*; or the frequent application and removal of stress of the same kind—was sufficient to fracture a rod, even if the maximum stress never exceeded one which was safe when the load was a permanent or steady dead load.

Effect of Temperature on Metals and Alloys.— The strength of steel and good wrought iron is not adversely affected by temperatures up to 500° F., but above this point the tenacity begins to fall off. Cast iron appears to be somewhat unreliable at temperatures below 32° F., and its strength begins to fall off at about 200° F. Brass and gunmetal are unsuited for temperatures above 400° F., and should not be used in valves and steam fittings when the steam is superheated. Special alloys suitable for fairly high temperatures can be obtained. The strength of copper falls off at a temperature of about 300° F.

Young's Modulus of Elasticity.—The student will frequently find references to Young's modulus. This modulus is the ratio that a given stress per unit of section bears to a given elongation within the elastic limit per unit of length. This coefficient of elasticity varies with the quality of the material; for good wrought iron it is about 12,000, for mild steel about 13,000, and for cast iron about 7,000, if tons per square inch are taken. Thus, if a force of 15 tons should stretch a bar 800 inches long and 1 square inch area to the extent of 1 inch, the material still retaining its elasticity, then the coefficient of the material would be 12,000.

Factors of Safety.—The factor of safety, or number by which the breaking strength of a material should be divided to give its safe strength, depends entirely upon the conditions under which the material will be used. In building construction where the load may be a permanent or dead one, a factor of safety as low as 3 is sometimes taken for steel work. In boiler work in this country a factor of safety of about 5 is usually taken; thus, if the breaking strength of the steel plate is 30 tons, the safe stress will be assumed to be 6 tons. In the case of iron castings where internal stresses may be set up in cooling, and in which there may be small blowholes, the factor of safety taken is usually not less than 8. In deciding upon the factor of safety, the results of Wöhler's investigations, previously referred to, must not be overlooked. Professor Unwin, who is an authority on testing and on the strength of materials, considers that Wöhler's experiments show roughly that if the safe stress of a steel bar under a steady permanent load is called 3, then the safe stress for the same bar under a load which is alternately removed and replaced, will be 2; while if the bar is subject alternately to tension and compression, the safe stress will be only 1.

Quality.—The following clauses dealing with the quality of cast iron, mild steel, and cast steel are taken from actual specifications issued by engineers within the last year or two:—

"**Cast Iron.**—All cast iron to be of good close-grained quality, free from cracks, flaws, blowholes, or chilled spots. All castings under working stresses to be of metal to stand the following tests, viz.:—A test bar 42 inches long by 2 inches by 1 inch in section to be cast at the same time and from the same ladle. The test bar when placed on its edge between supports 36 inches apart to carry a load (gradually applied) in the centre, of 30 cwts., and to deflect under such load ·2 inch. The harder metal for the cylinders and liners to carry a load under the same conditions of 40 cwts., and to deflect under such load ·3 inch.

"**Mild Steel.**—The crank-shaft, piston-rods, and connecting-rods to be of mild steel, having a tenacity of from 26 to 30 tons per square inch, and with an elongation of not less than 22 per cent. in 10 inches before fracture. If required to do so by the engineers, the contractors must send them ready prepared specimens (of dimensions approved by them) cut from the forgings for the purpose of being tested.

"**Mild Steel Boiler Plates.**—All the plates throughout each boiler, and all the rivets and gusset stays are to be made of open hearth mild steel. All the plates are to be placed with their brands visible on the outside of the shell, or on the inside of the flues, as the case may be. Every plate throughout the boiler is to be tested at the expense of the contractor under this specification by a longitudinal and transverse strip taken from each plate. Every plate to be capable of standing a tensile stress of somewhere between the limits of 27 tons and of 30 tons per square inch of original sectional area, either lengthwise or crosswise of the plates, with an extension of not less than 20 per cent. in a length of 10 inches, which is to be the length of the operative part of the test piece. In no case is the test piece to have a less sectional area than half a square inch.

"**Cast Steel.**—All cast steel to be as free from blowholes as practicable, and to be well annealed. All large blowholes to be filled by electric welding, and castings so treated to be annealed after welding. Test bars to be cast from the same ladle at the same time as the bulk of the castings, and to give a tensile strength of 27 to 30 tons per square inch, and an elongation in 3-inch test bars of not less than 20 per cent. before fracture.

"**Manganese Bronze.**—All manganese bronze used for bolts, nuts, studs, &c., to be of high tensile strength, and test bars to show a tensile strength of 25 to 28 tons per square inch, and an elongation of not less than 20 per cent. before fracture."

Boiler makers, in ordering steel plates from the makers, usually specify that a strip 2 inches broad and 10 inches long, sheared from the plate and heated to redness, then cooled in water at 80° F., must stand bending until the inside radius of the curve is one and a-half times the thickness of the plate. The plates must stand this bending without showing any signs of fracture.

Table III., giving the approximate cost of the material with which an engineer has to deal, may prove useful to a beginner.

TABLE III.—Approximate Cost of Materials.

Iron castings, plain and fairly heavy,	£9 to £12	per ton.
,, intricate ,,	£12 ,, £18	,,
Pig iron (Scotch),	£2 15s. ,, £3 10s.	,,
Wrought iron, or Mild steel, Plates, angles, and bars,	£6 ,, £8 10s.	,,
Steel castings,	£28 ,, £45	,,
Aluminium ingots,	£120 ,, £130	,,
Gunmetal castings,	10d. ,, 1s. 6d.	per lb.
Phosphor bronze rods,	10d. ,, 1s.	,,
Manganese ,,	11d. ,, 1s.	,,
Stone's ,,	11d. ,, 1s.	,,
Copper ingots for melting,	£70 ,, £115	per ton.
Tin ,, ,,	£150 ,, £200	,,
Zinc ,, ,,	£170 ,, £195	,,
Lead ,, ,,	£17 ,, £22	,,
White metal ,,	£50 ,, £200	,,

N.B.—The wide margin given in the prices of tin and copper is due to the fact that the price of these metals fluctuates very considerably.

CHAPTER II.

BOLTS AND NUTS, STUDS, SET SCREWS AND RIVETS.

BOLTS and nuts, studs, set screws and rivets are made of wrought iron or mild steel; the annexed illustrations clearly show the difference between each. Bolts, studs, and set screws are used for bolting together two pieces of metal which may require subsequently to be taken apart. Bolts and nuts are used in all cases where there is room for the bolt head. Studs are used where there is not room for a bolt head, as shown by Fig. 4, or where it is undesirable to make a hole right through both pieces of metal to be fastened together. The objection to a stud is that should it break off, it is difficult to extract the portion of the stud which is screwed firmly into the metal. Even if the thread only gets stripped off, or the stud becomes bent, it is a troublesome matter to replace it. A set screw is used in cases where there is not room for a bolt head, and where it is undesirable to have a projecting stud when one portion of the joint has been removed.

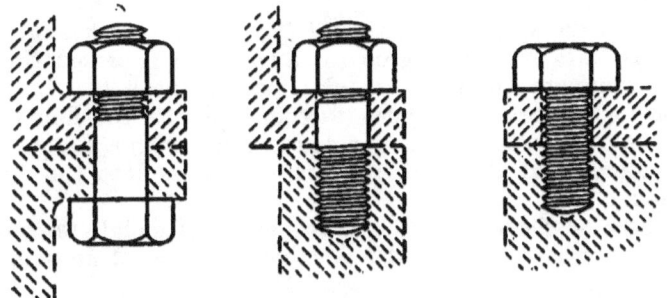

Fig. 3.—Bolt and nut.   Fig. 4.—Stud.   Fig. 5.—Set screw.

In cases where bolts and nuts are subject to vibration, there is a danger of the nuts working loose and coming off. Various devices are employed to minimise this danger: the oldest and most common one is to use two nuts and to lock them together. The two nuts are screwed down hard, the inner nut is then unscrewed for a fraction of a turn, the outer nut being screwed down at the same time. The nuts are thus pressed one against the other and are locked; two spanners are required for the operation. A split pin is usually put through the end of the bolt to prevent any chance of the nuts coming off, should they come unlocked.

Fig. 6. Lock nuts.   Fig. 7. Grover washer.   Fig. 8. Castle nut.

Another device recently introduced is the Grover washer, which consists of a split spring washer, as shown in Fig. 7. When the nut is screwed down it compresses the washer, and thus the threads of the nut are always pressed tightly against those of the bolt. Grover washers of small size are of plain rectangular section and not as shown by the illustration. In the Helicoid nut (Fig. 9) the nut itself takes the form of a spring and grips the bolt.

Fig. 9.—Helicoid nut.

Another device is the Castle nut (Fig. 8); this nut is provided with saw cuts or narrow grooves; a split pin is passed through one of these grooves and through a hole in the bolt. If one of the saw cuts does not happen to come opposite the hole in the bolt, the nut is removed and a few touches with a file given to the underside until one of the grooves comes into the desired position.

Steady or Dowel Pins.—If two pieces of metal are fastened together by set screws and are subject to lateral motion, such

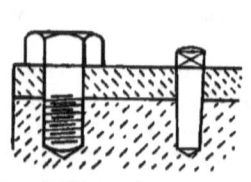

Fig. 10.—Set screw and steady pin.   Fig. 11.—Lewis bolt.

motion causes the sharp edges of the threads to cut into the adjoining metal; to overcome this, one or more steady or dowel pins are usually put in as shown by Fig. 10. A steady pin is merely a plain round piece of steel slightly tapered and driven into a correspondingly tapered hole; it effectually prevents any lateral motion.

Fig. 11 shows a Lewis bolt, such as is sometimes used for holding down machine-tools or small pieces of machinery to concrete foundations. The Lewis bolts are put into holes large enough to receive the largest part of the bolt; the space round the bolt is then run in with cement or molten lead. Large pieces of machinery usually have long holding bolts provided with square flat anchor plates at the lower end. In some cases the bolt has a solid head under the anchor plate, in which case the chase or square hole containing the bolt and anchor plate is not run in with cement until the bed-plate has been placed in

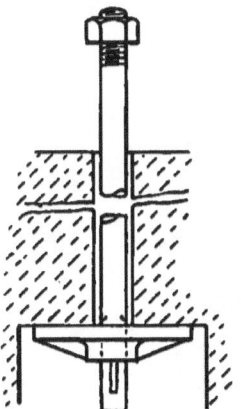

Fig. 12.—Foundation bolt and anchor plate.

position. In other cases there is an oblong hole in the anchor plate through which the bolt-head, also oblong, is dropped. The anchor plate is provided with stops which prevent the bolt-head making more than half a turn—*i.e.*, until the oblong bolt-head is at right angles to the oblong hole. A bolt with a head and plate of this description can be withdrawn and replaced should it be necessary to do so. The hole through the foundation is, of course, left sufficiently large for the head to pass.

In other cases hand holes are provided in the foundations to give access to the lower ends of the bolts, and a cotter or flat piece of steel, as shown by Fig. 12, is placed through a slot in the bolt, and serves as a head. If it is desired to replace the bolt, the cotter is knocked out and the bolt drawn up through the foundation. The end of the bolt passing through the anchor plate is square. This gives increased section to compensate for the slot, and prevents the bolt from turning while the nut is being tightened up.

The following table, giving the number of threads per inch of a Whitworth screw, and the diameter of the bolt at the bottom of the thread, with the corresponding area, may be found useful :—

TABLE IV.—WHITWORTH THREADS.

Diameter of Bolt.	Number of Threads per Inch.	Diameter at Bottom of Thread.	Area at Bottom of Thread.	Thickness of Head.	Thickness of Nut = Diameter of Bolt.
Inches.		Inches.	Sq. Inches.	Inches.	Inches.
$\frac{1}{4}$	20	·186	·027	·219	$\frac{1}{4}$
$\frac{3}{8}$	16	·295	·068	·328	$\frac{3}{8}$
$\frac{1}{2}$	12	·392	·121	·437	$\frac{1}{2}$
$\frac{5}{8}$	11	·508	·202	·547	$\frac{5}{8}$
$\frac{3}{4}$	10	·622	·303	·656	$\frac{3}{4}$
1	8	·840	·553	·875	1
$1\frac{1}{4}$	7	1·067	·894	1·092	$1\frac{1}{4}$
$1\frac{1}{2}$	6	1·286	1·298	1·312	$1\frac{1}{2}$
$1\frac{3}{4}$	5	1·493	1·750	1·531	$1\frac{3}{4}$
2	$4\frac{1}{2}$	1·715	2·309	1·75	2
$2\frac{1}{2}$	4	2·179	3·724	2·18	$2\frac{1}{2}$
3	$3\frac{1}{2}$	2·634	5·439	2·62	3
4	3	3·573	10·027	3·5	4
5	$2\frac{3}{4}$	4·534	16·146	4·5	5
6	$2\frac{1}{2}$	5·489	23·65	5·25	6

The ordinary Whitworth thread is rather coarse, and cuts into the metal to a considerable extent. Wrought-iron pipes for steam, water, and gas are screwed with a much finer thread, which is usually known as "gas thread." This thread is often used for parts of machinery where a fine adjustment is required, and where it is inadvisable to cut deeply into the metal. In speaking of a $\frac{1}{4}$-inch gas thread, a thread suitable for a pipe, the internal diameter of which is $\frac{1}{4}$ inch, is meant. The external diameter, the number of threads per inch, and the diameter at the bottom of the thread, are given in Table v.

To ascertain what load a rod of wrought iron, say 1 inch in diameter, screwed with Whitworth thread will safely carry, it is necessary to take the area of the bolt at the smallest part—viz., at the bottom of the thread. From Table IV. it will be seen

TABLE V.—Gas Threads.

Internal Diameter of Pipe.	Number of Threads per Inch.	Diameter Outside.	Diameter at Bottom of Thread.
Inches.		Inches.	Inches.
1/8	28	·382	·336
1/4	19	·518	·456
3/8	19	·656	·588
1/2	14	·825	·734
5/8	14	·902	·810
3/4	14	1·041	·949
7/8	14	1·189	1·097
1	11	1·309	1·192
1 1/8	11	1·492	1·375
1 1/4	11	1·650	1·523
1 3/8	11	1·745	1·628
1 1/2	11	1·882	1·766
1 5/8	11	2·021	1·904
1 3/4	11	2·116	2·000
2	11	2·347	2·231

that the area of a 1-inch bolt at the bottom of the thread is only ·553 of an inch. If, however, it is decided to use a gas thread, so as not to cut so deeply into the metal, and we wish to know the area at the bottom of the thread, we must refer to Table v. From this it will be seen that the gas thread of a 3/4 pipe is 1·041 inches outside, so that a 3/4 gas thread would be used. The diameter at the bottom of this thread is ·949, as compared with ·840 in the case of the Whitworth thread.

In calculating the load a bolt will bear, it must not be overlooked that a considerable stress may be imparted to the bolt by the act of tightening up the nut, and for this reason a considerable margin must be allowed. It is not good practice to allow a greater stress than 1 ton per square inch on bolts under 3/4-inch diameter, or 1 1/2 to 2 tons per square inch upon bolts above this size. Bolts, the nuts of which require to be well tightened up, should be not less than 5/8-inch in diameter. With ordinary force applied at the end of a spanner it is possible to break a 3/8-inch bolt, and to put an undue stress upon a 1/2-inch bolt. Calculation shows that, neglecting friction, a force of 10 lbs. applied at the end of a spanner 8 inches long is sufficient to cause a stress of 53 tons per square inch on a 3/8-inch bolt, and 22 tons per square inch on a 1/2-inch bolt. In actual practice, however, the friction between the nut and the metal on which it presses absorbs some of the force applied to the spanner. The friction between the threads of the nut and bolt transmits a torsional stress to the latter.

As the diameter of a bolt increases, the area of metal increases very rapidly—viz., as the square of the diameter, and the stress per square inch due to tightening up falls off. In the case of a 1-inch bolt and a force of 20 lbs. applied at the end of a spanner, 10 inches long, a stress of about 8 tons per square inch is put on the bolt. We see from the table that the area of a 1-inch bolt is ·553 inch at the bottom of the thread, and if we allow a stress of about $1\frac{1}{2}$ tons per square inch, we find that the bolt will carry ·83 of a ton in addition to the stress put upon it by tightening up the nut, or 8·83 tons per square inch altogether. This, although a fairly high stress, is well below the elastic limit of wrought iron, and is only about one-third of the breaking stress of the metal.

The most suitable number of bolts, their diameter, and pitch, for circular flanges will be found in the table of pipe flanges, prepared by the Engineering Standards Committee, given in Chapter v. It will be noticed that bolts smaller than $\frac{5}{8}$-inch in diameter are not used, even with the smallest flanges.

Rivets are only used for wrought iron and steel work, and in cases where the two pieces of metal will not require to be taken apart at any future time. Rivets, unless of very small size, are heated before being closed; the contraction of the rivet while cooling draws the plates tightly together. Large rivets are usually closed by means of hydraulic riveters; such riveting machines are described in the chapter on hydraulic machinery.

Figs. 13 and 14.—Rivet before and after closing.

The size of rivets and the distance they are spaced apart or pitched depends upon the thickness of the plate and the kind of joint used. The following tables give approximately the dimensions found in practice :—

TABLE VI.

SINGLE-RIVETED LAP JOINT.			
Thickness of Plate.	Diameter of Rivet.	Pitch of Rivet.	Lap of Plates.
Inches.	Inches.	Inches.	Inches.
$\frac{3}{8}$	$\frac{3}{4}$	$1\frac{3}{4}$	2
$\frac{7}{16}$	$\frac{13}{16}$	2	$2\frac{1}{4}$
$\frac{1}{2}$	$\frac{7}{8}$	$2\frac{1}{8}$	$2\frac{5}{8}$
$\frac{9}{16}$	$\frac{15}{16}$	$2\frac{1}{4}$	$2\frac{3}{4}$
$\frac{5}{8}$	1	$2\frac{1}{4}$	3
$\frac{11}{16}$	$1\frac{1}{8}$	$2\frac{1}{2}$	$3\frac{1}{4}$
$\frac{3}{4}$	$1\frac{1}{8}$	$2\frac{1}{2}$	$3\frac{1}{4}$

Double-riveted Lap Joint.

Thickness of Plate.	Diameter of Rivet.	Pitch of Rivet.	Lap of Plates.
Inches.	Inches.	Inches.	Inches.
$\frac{1}{2}$	$\frac{13}{16}$	$2\frac{1}{2}$	$3\frac{3}{4}$
$\frac{9}{16}$	$\frac{13}{16}$	3	$4\frac{1}{4}$
$\frac{5}{8}$	$\frac{7}{8}$	3	$4\frac{1}{4}$
$\frac{11}{16}$	1	$3\frac{3}{8}$	$4\frac{1}{2}$
$\frac{3}{4}$	$1\frac{1}{16}$	$3\frac{1}{2}$	$4\frac{3}{4}$
$\frac{7}{8}$	$1\frac{1}{8}$	$3\frac{1}{2}$	$5\frac{1}{4}$
1	$1\frac{5}{16}$	4	$5\frac{1}{2}$

Double-riveted Butt Joints — Double Straps.

Thickness of Plate.	Diameter of Rivet.	Pitch of Rivet.	Width of Butt Straps.
Inches.	Inches.	Inches.	Inches.
$\frac{1}{2}$	$\frac{13}{16}$	$3\frac{1}{4}$	$8\frac{1}{2}$
$\frac{5}{8}$	$\frac{7}{8}$	$3\frac{1}{2}$	$8\frac{3}{4}$
$\frac{3}{4}$	$\frac{7}{8}$	$3\frac{1}{2}$	9
$\frac{7}{8}$	1	4	10
1	$1\frac{1}{8}$	$4\frac{1}{2}$	11
$1\frac{1}{8}$	$1\frac{1}{4}$	$5\frac{1}{4}$	12
$1\frac{1}{4}$	$1\frac{5}{16}$	$5\frac{1}{2}$	13

Treble-riveted Butt Joints — Double Straps.

Thickness of Plate.	Diameter of Rivet.	Pitch of Rivet.	Width of Butt Straps.
Inches.	Inches.	Inches.	Inches.
$\frac{3}{4}$	$1\frac{3}{16}$	$3\frac{3}{4}$	12
$\frac{7}{8}$	$1\frac{3}{16}$	$4\frac{1}{2}$	$13\frac{1}{2}$
1	$1\frac{5}{16}$	6	18
$1\frac{1}{8}$	$1\frac{5}{16}$	$6\frac{5}{16}$	$19\frac{3}{4}$
$1\frac{1}{4}$	$1\frac{3}{8}$	$6\frac{5}{8}$	21
$1\frac{3}{8}$	$1\frac{5}{8}$	$7\frac{3}{4}$	24

A lap joint, as its name implies, is one in which one plate laps over the other. A butt joint is one in which the two edges of the plate meet; separate straps or cover plates are used on one or both sides of the plate.

The pitch referred to in the tables is the distance from centre to centre of the rivets in the rows running parallel to the joint. In double riveting the diagonal pitch—*i.e.*, the distance from the centre of the rivet in one row to the centre of the rivet in the next row, measured diagonally, should be not less than $\cdot 65 \ P + \cdot 35 \ D$; where P = pitch of rivets, D = diameter of rivets.

The thickness of conical rivet heads should not be less than three-quarters of the diameter of the rivet. The strength of riveted joints is dealt with in the chapter on boilers.

CHAPTER III.

BOILERS.

BOILERS may be divided into two distinct classes—viz., the Shell and the Water-tube. Boilers of the Cornish, Lancashire, locomotive, and kindred types are shell boilers—*i.e.*, they consist of a cylindrical shell containing the water and steam. The shell contains combustion chambers in which the fuel is burnt, and flues or tubes through which the gases pass; the combustion chamber and tubes are surrounded by water.

In the water-tube class of boiler the water is contained in, and circulated through, a large number of comparatively small tubes; the tubes are connected either directly to, or through headers with, drums containing steam and water. Amongst the best known water-tube boilers are the Babcock, Stirling, Thornycroft, Yarrow; there are many others. We will first consider boilers of the shell type, pointing out their good features and the dangers connected with their use.

Lancashire Boiler.—Fig. 15 shows a Lancashire boiler in section, and Fig. 16, a front view of the same boiler. This type of boiler has been in use for a great many years, and has many points in its favour. It is fairly inexpensive to construct, and is sufficiently large to allow of a man getting inside to inspect and to chip away any scale, should such be formed. When steam has been raised it is easy to maintain it at a constant pressure, even when sudden demands for large quantities of steam are made upon it, owing to the large volume of water (ready to evaporate) and of steam which it contains. The objections to this boiler are those which apply to all boilers of the shell type, and are dealt with later. Apart from these objections, all that can be urged against a boiler of the Lancashire type is that it occupies a good deal of space, and a considerable time is required to raise steam if the boiler is allowed to get cold.

Fig. 15 represents a 30 feet by 8 feet Lancashire boiler suitable for a pressure of 160 lbs.; there are a few points about it to which attention should be drawn.

In the first place, a true section through the centre of the boiler would pass between the two flues, and the outside of one flue only would be seen. In order to show the inside of the

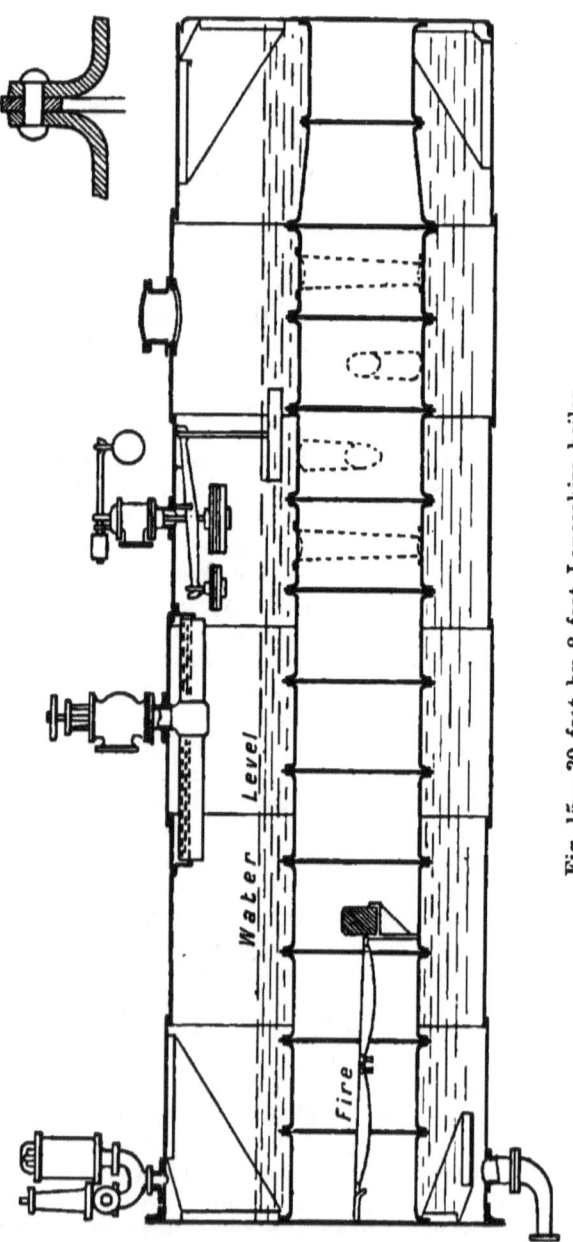

Fig. 15.—30 feet by 8 feet Lancashire boiler.

flue, a section following the dotted line A A, B B (Fig. 16), is given. This conventional way of showing a Lancashire boiler has some drawbacks—for instance, the gusset stays, which prevent the ends from bulging out under the pressure of steam, appear to go close up to the flue, while in reality there is a space of 9 inches or 10 inches between the stay and the flue, as will be seen from the end view. The object of this breathing space, as it is called, is to allow the end plates to "give" slightly when the flues expand or contract.

It will be noticed that while the shell consists of six rings of rather wide plates, the flue is composed of a much larger number

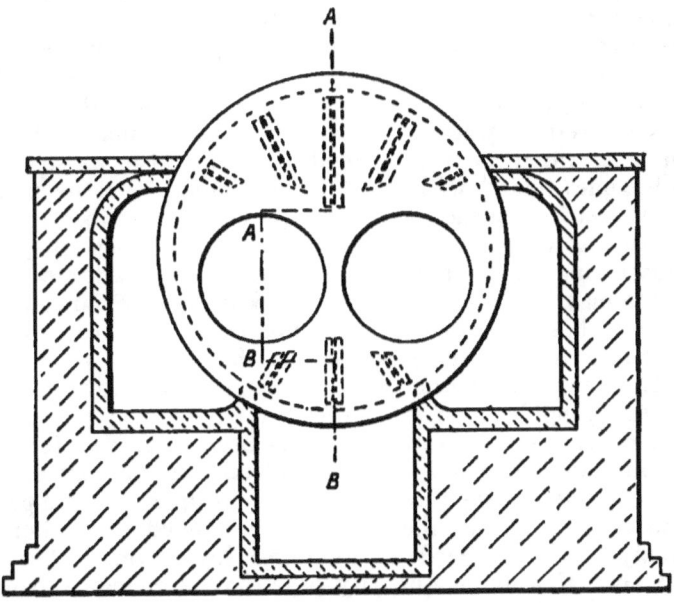

Fig. 16.—Lancashire boiler, with brick setting.

of narrow rings. The reason for this is as follows :—The fewer plates there are forming the shell the less riveting is required, and a sounder and less expensive boiler is produced. As much wider plates can be obtained now than was formerly the case, the number of rings forming the shells of Lancashire and other boilers has been considerably reduced. The width of the furnace rings, when joined together by "Adamson" rings, is determined, on the other hand, by the working pressure of the boiler, as the strength of the flue to resist collapsing pressure is largely dependent upon the number and distance apart of these rings.

In a boiler constructed for only 80 lbs. working pressure, the furnace rings may be about 3 feet 6 inches wide, and the Adamson rings be spaced this distance apart from one another; but for a boiler to work at a pressure of 160 lbs., the furnace rings should only be 2 feet 3 inches wide; while for a pressure of 200 lbs. the rings should be still narrower—viz., about 2 feet 1 inch wide.

An enlarged sectional view of the Adamson ring is given in the top right-hand corner of Fig. 15. It will be noticed that the rivet heads are not exposed to the flames or gases, while the central plate enables the joint to be caulked.

The figure shows four Galloway cross tubes in dotted lines only, as the modern tendency is to dispense with them. These tubes give increased heating surface, and for many years were believed to improve the circulation of water in a boiler, but the manner in which the tubes become coated with scale has made some engineers sceptical as to whether the circulation through them is as rapid as was thought. In any case, as soon as a tube becomes coated with scale, its efficiency, from the point of view of heating surface, rapidly falls off. The tubes act very efficiently as cross struts to support the flue, but by placing Adamson rings closer together, as already described, cross tubes as struts may be dispensed with. Leakage at the flanged joints of these tubes is also sometimes experienced, and to avoid this some boiler makers have welded in the cross tubes, but this operation is not altogether an easy one, and, unless the welding is very thoroughly done, it may lead to trouble.

The old-fashioned steam dome has been discarded for many years on Lancashire and Cornish boilers. The reasons are twofold; in the first place, the dome did not ensure dry steam passing to the engine, and in the second it weakened the boiler considerably. With regard to obtaining dry steam, it has been found that if a large quantity of steam is collected from an aperture placed over a small area of water, the rush of steam carries with it small particles of water, and this is called "priming." An anti-priming pipe, as shown by Fig. 15, is now generally fitted; this is merely a pipe with closed ends, having a large number of slots or perforations in the upper half of its circumference. This pipe collects steam evenly from a fairly large area, and is more effective than the old-fashioned steam dome.

The fitting shown at the extreme left of the boiler consists of one spring loaded and one dead weight safety valve mounted on a common seating. The spring safety valve is usually provided with a lever, by moving which (by a chain or otherwise) the valve may be raised and steam blown off through a pipe fitted to it. The dead weight safety valve is usually loaded to blow off at 5 lbs. greater pressure than the spring valve, the steam escaping to the boiler-house.

The fitting shown to the right of the stop valve and anti-priming pipe is a high pressure and low water alarm safety valve. If the level of the water falls too low the float sinks and opens a small valve which may communicate with a whistle; in the figure the float is apparently almost touching the flue, because the figure does not show a true section; in reality the float can descend for a short distance between the two flues. Surrounding the small valve, actuated by the float, there is another valve which acts as an ordinary safety valve should the steam pressure rise too high.

The manhole shown to the right of the low water alarm is to enable a man or boy to get inside the boiler for examination or for cleaning: a strengthening ring is riveted round the shell, just below the manhole.

The elbow at the bottom of the boiler is for blowing out any sediment which may have accumulated. The elbow is usually provided with a cock; this cock is one of the most troublesome fittings about a boiler, as it gets cut by the outrushing steam, water, and dirt. Some users employ a cock next the boiler with a valve beyond it; by keeping the latter closed until the cock is fully opened, the cutting action which is supposed to occur while the cock is being opened is thus minimised. Any slight leakage of the outer valve is immaterial.

The smaller fittings, such as the steam pressure gauge for showing the pressure of steam, and the water gauge fittings for showing the level of the water in the boiler, are not shown. The latter are usually provided with thick glass shields or protectors, or the glass tubes are partly surrounded by a brass casing, so that in the event of the glass breaking, the pieces may not strike the stoker in the face.

The following table, giving the sizes and approximate evaporation of some Lancashire boilers of the usual sizes, may be found useful:—

TABLE VII.—Lancashire Boiler.

Size of Boiler.				Diameter of Flue.		Grate Area.	Heating Surface.	Approximate Evaporation.
Ft.	Ins.	Ft.	Ins.	Ft.	Ins.	Sq. Ft.	Sq. Ft.	Lbs. per hour.
20	0 ×	6	6	2	7	20	507	4,250
22	0 ×	6	6	2	7	22	563	4,700
24	0 ×	6	6	2	7	24	620	5,200
24	0 ×	7	0	2	9	25	670	5,600
26	0 ×	7	0	2	9	27	730	6,000
28	0 ×	7	0	2	9	30	790	6,500
30	0 ×	7	0	2	9	33	850	7,000
30	0 ×	7	6	3	0	36	890	7,500
28	0 ×	8	0	3	2	36	920	7,500
30	0 ×	8	0	3	2	38	960	8,000
30	0 ×	8	6	3	4	40	1,010	8,500

Note.—The evaporation given for each boiler is "from and at 212° F.,"* and is based on the assumption that between 20 and 25 lbs. of coal are burnt per square foot of grate per hour. With a mechanical stoker more coal than this can be burnt and greater evaporation obtained; with hand firing, unless the fireman is very capable, less fuel will probably be burnt and a smaller evaporation obtained.

Cornish Boiler.—This boiler is similar to the Lancashire, but is smaller, and has only one internal flue instead of two; a front view of this boiler is shown by Fig. 17. A Cornish boiler

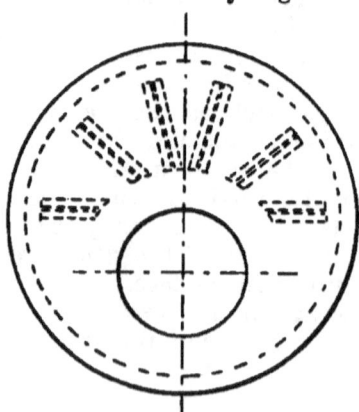

Fig. 17.—Cornish boiler.

is suitable in cases where an evaporation of 1,000 to 4,000 lbs. of water is required per hour. The following is a list of the sizes usually made, with the approximate evaporation which may be expected:—

* The explanation of this expression is given later.

TABLE VIII.—Cornish Boiler.

Size of Boiler.				Diameter of Flue.		Heating Surface.	Approximate Evaporation.
Ft.	Ins.	Ft.	Ins.	Ft.	Ins.	Sq. Ft.	Lbs. per hour.
14	0 ×	5	0	2	8	220	1,250 to 1,700
16	0 ×	5	0	2	8	250	1,400 ,, 1,900
18	0 ×	5	0	2	8	280	1,600 ,, 2,100
20	0 ×	5	0	2	8	310	1,750 ,, 2,300
18	0 ×	6	0	3	2	340	1,950 ,, 2,600
20	0 ×	6	0	3	2	380	2,200 ,, 2,900
22	0 ×	6	0	3	2	400	2,400 ,, 3,200
24	0 ×	6	0	3	2	440	2,600 ,, 3,500
26	0 ×	6	6	3	3	530	3,000 ,, 4,000

In the majority of places where Cornish boilers are installed, the stoker has duties other than those connected with the boiler; under these circumstances a boiler which is somewhat large for the output required is usually preferred, so that the stoking may be done intermittently, without causing any serious fluctuations in the steam pressure.

A **Galloway** boiler is a modification of the Lancashire; the two tubes are merged into one for a great part of their length, and a large number of cross tubes are fitted. It is claimed that a Galloway boiler will evaporate more water than a Lancashire boiler of the same external dimensions, but, the form of the flue not being truly cylindrical, it is not considered so suitable for withstanding high pressures as the flue of a Lancashire boiler.

The **Economic** boiler (manufactured by Davey, Paxman & Co.) is similar to a Cornish boiler, but is provided with a number of small tubes running parallel to the main flue, through which the hot gases return. This boiler occupies less space than the Cornish, but more time requires to be spent upon it in cleaning the tubes and keeping them tight. The boilers just described are set in brickwork, flues being formed in it to bring the hot gases back alongside the outside of the shell; these gases then return underneath the boiler to the chimney.

Fig. 18 shows a **Locomotive** boiler. This type of boiler is employed in cases where a large amount of steam is required from a comparatively small boiler, where portability is required, and where brick setting is not admissible. It is used, as its name implies, in locomotives, also for portable engines, traction engines, road rollers, &c.; it is occasionally used in steam installations of a temporary nature, and on some torpedo boats. When used on a locomotive the boiler is capable of generating

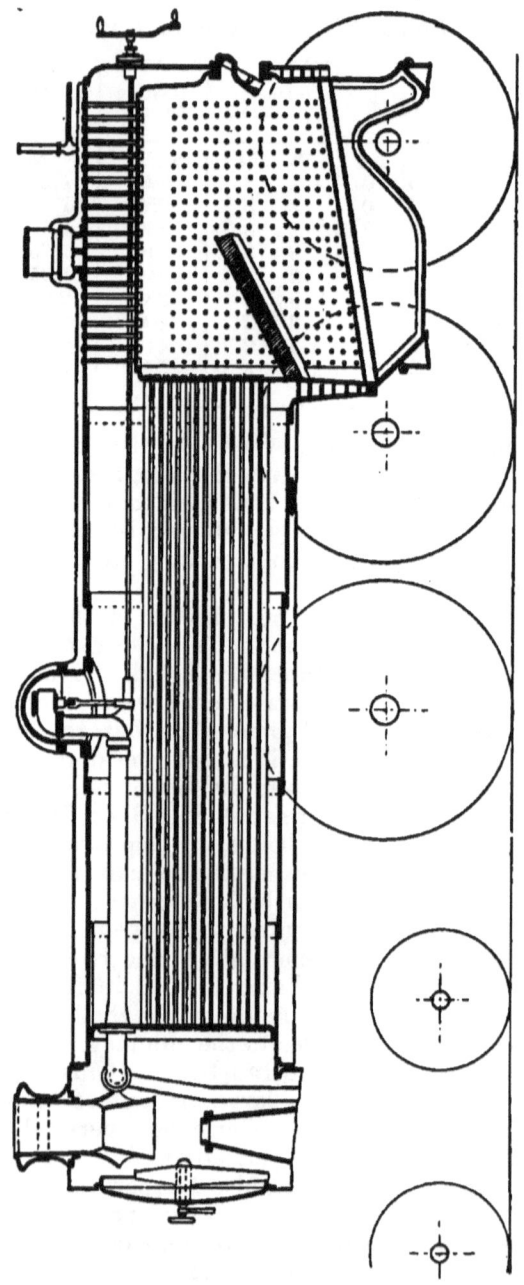

Fig. 18.—Locomotive boiler.

BOILERS.

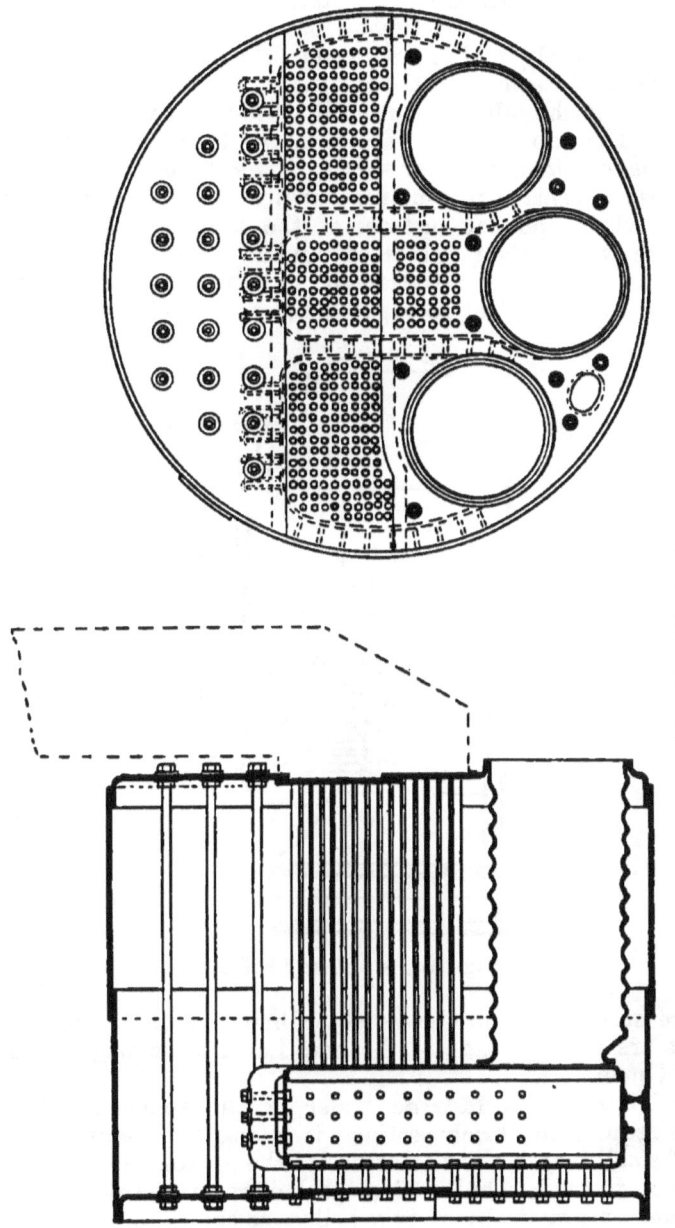

Fig. 19.—Marine boiler.

more steam than when used in a stationary position; this is due partly to the powerful draught caused by the blast of exhaust steam, and partly to the vibration which is believed to free the bubbles of steam from the tubes.

In an English-built locomotive the fire-box is usually made of copper. The tubes, which vary from $1\frac{3}{4}$ to 2 inches diameter, are made of steel, iron, or brass. The locomotive boiler is somewhat expensive to construct, owing to the large amount of staying which the flat surfaces of the fire-box require; these stays are frequently a source of trouble, owing to leakage and breakages. A recent innovation on the Lancashire and Yorkshire Railway is a mild steel corrugated fire-box which is said not to

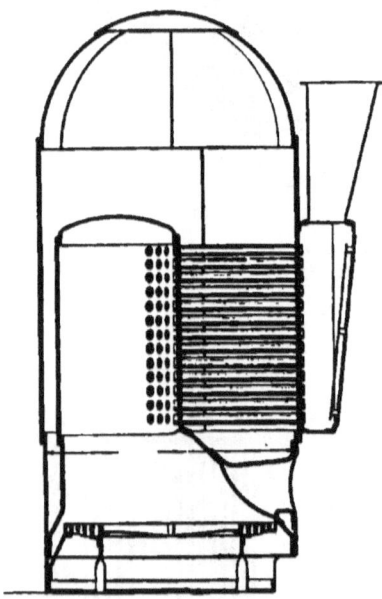

Fig. 19a.—" Blake " boiler.

require stays. The fire-boxes of ordinary portable (loco. type) boilers are usually made of the best Lowmoor wrought iron, or of mild steel.

Fig. 19 shows a **Marine** boiler of the single-ended type, which requires no brick setting; it is made of large diameter and of small length, so as to utilise to the greatest advantage the shape of the ship's hull. Marine boilers of the double-ended type consist practically of two boilers as shown, placed back to back, but with one combustion chamber common to both.

A **Dry Back** boiler is similar to the single-ended marine, but the back of the combustion chamber is formed of brickwork.

The marine boiler is fairly satisfactory in use; its disadvantages are—the great weight of the boiler when full of water, and the length of time required to get up steam. The stays in the combustion chamber sometimes give trouble, and the tubes require to be expanded occasionally.

Vertical Boilers. — A large variety of vertical boilers requiring no brickwork are made, and are suitable for cases where a small quantity of steam only is required. The boiler consists of a vertical shell, having an internal fire-box; there are either cross tubes upon which the flames impinge, or a number of small tubes through which the gases pass on their way to the chimney. These boilers are compact and handy, but are not economical in fuel, as a large percentage of the heat generated passes away through the chimney.

One of the most economical boilers of this type is the Blake (Fig. 19a), made by The Blake Engineering Co., Ltd., Darlington. This boiler has no flat surfaces and no stays.

Dangers of the Shell Boiler.—Unless boilers of the shell type are examined periodically their use is attended with serious risk of explosion, due to the plates becoming weakened, either by grooving, pitting, or corrosion. Owners of such boilers usually insure them against explosion, and the Boiler Insurance Co. periodically sends specially qualified men to examine and report as to the condition of the boilers. Grooving consists of the formation of grooves in the plates of a boiler, usually near a joint; the grooves are sometimes deep and narrow, and sometimes wide and shallow. They are thought to be caused by undue stresses coming upon a small area of the boiler plates through unequal expansion and contraction. When the fire is first lighted in a Cornish or Lancashire boiler the temperature of the flues is raised, the outer shell remaining cool, and the expansion of the flue, which in a 30-feet boiler may be as much as $\frac{3}{8}$ inch, tends to force out the end plates. The end plates, except in the Thomson boiler, referred to later, are held in by gusset stays, and thus, if the plates are thick and unyielding, undue stresses may come upon portions of them.

It was found some years ago by Mr. Andrews, who made a study of the corrosion of metals, that when two pieces of steel, one of which had been strained to breaking point, and the other subjected to very slight stress, were placed in a salt solution, they formed a galvanic couple, and that the galvanic action, thus set up, greatly increased the corrosion.

38 MECHANICAL ENGINEERING FOR BEGINNERS.

Pitting, or the formation of groups of small holes in a boiler plate, is usually due, in the first place, to the presence of some acid, or impurity in the feed water, which attacks the plates. Wasting is the gradual decrease in thickness of the plates, due to corrosion.

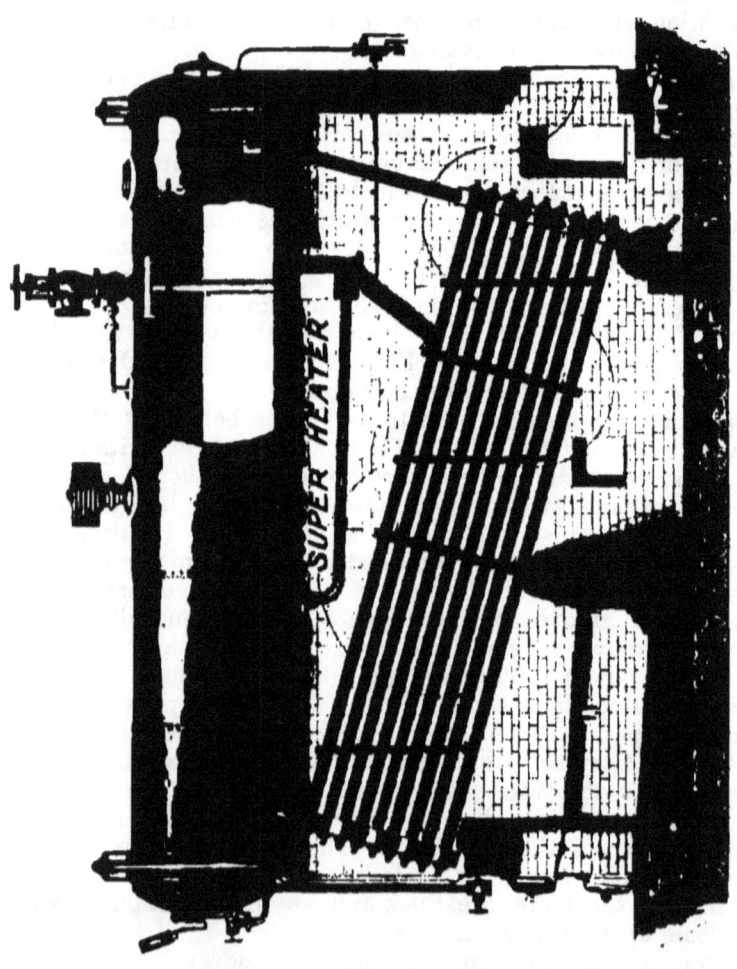

Fig. 20.—Babcock boiler.

To avoid the stresses set up by the expansion of the flues, a portion of the flue is sometimes made corrugated, as shown in Fig. 19. It will be noticed in the illustration of the Lancashire boiler that the end plates are stayed, or prevented from being forced outwards by the pressure of the steam, by gusset stays. The latest development in Cornish and Lancashire boilers in this country consists of dishing the ends of the boiler outwards in the form of a saucer; these ends being partly spherical, are said not to require stays, and thus one of the sources of undue stress in a boiler is removed. These dished boiler ends are, however, stiff, and some engineers consider that corrugated flues should be used when dished ends are employed. The dished boiler ends have recently been introduced by Messrs. Thomson & Co., Wolverhampton.

The plates of shell boilers are almost invariably made of mild steel. The rivet holes are all drilled, and the rivets are of mild steel.

WATER-TUBE TYPE.

We will now consider boilers of the water-tube type. Fig. 20 shows the **Babcock** boiler. This boiler consists of a large number of tubes, each about 4 inches external diameter, inclined at an angle and expanded at each end into vertical rectanglar tubes or headers; these headers communicate with a drum which is half full of water, the remainder of the drum forming a space for steam. The water descends by the back headers, rises through the inclined tubes, and passes up the front headers, thus maintaining a very good circulation. The furnace is placed under the front end of the tubes; the gases are deflected by firebricks, so that they pass completely over and under the whole length of the tubes, the gases striking them at right angles. The boiler, with the exception of the drum, is surrounded by brickwork, but it is slung from joists carried on columns, so that it is free to expand and contract.

The principal advantages of the Babcock boiler are as follows:—

Freedom from risk of explosion.—The steam and water drum is of small diameter, and has no flat stayed surfaces; it is not exposed to the fierce heat of the furnace, and is not subject to severe stresses owing to unequal contraction and expansion. The drum can therefore be made exceedingly strong for the pressure it has to sustain. The tubes, which are of comparatively small size, are capable of withstanding extremely high pressures; even in the rare event of a tube failing, the result is not very serious; the tube may be rent, but unless the fire door should happen to be open at the time (in which case the stoker might be scalded), no harm is done, and the tube can be renewed at a very small expense. Compared with the destruction brought about by an explosion of a boiler of the shell type, the result of a burst tube is insignificant.

Ability to raise steam quickly.—The water in a water-tube boiler is very much subdivided, and as the circulation is extremely good, steam can be raised very quickly; this is an important point in electric generating stations in towns where a fog may come on suddenly.

The boiler occupies a small space; the heating surface is at right angles to the path of the gases, and is thus in the best position for extracting the heat from them. The thickness of the metal through which the heat has to be transmitted is much less than in a boiler of the shell type. For marine work the Babcock boiler is slightly modified; smaller tubes are used and a metal casing, lined with specially light fire-bricks, takes the place of brickwork setting. Boilers of the water-tube type have a great advantage over shell boilers for war vessels, owing to their ability to raise steam quickly. Their light weight, due to the small amount of water held, is a point greatly in their favour. The disadvantages of this type of boiler are dealt with further on.

The overall length of a Babcock boiler is usually about 23 feet, irrespective of its evaporative capacity, but shorter boilers are made if required. The width may vary from 6 feet to 12 feet. The evaporation ranges from 3,000 to 20,000 lbs. per hour per boiler.

Niclausse Boiler.—This boiler is somewhat similar to the Babcock, but the tubes are connected to a header at one end only, by means of coned joints, the other end of the tube being closed. The circulation is obtained by placing one tube inside another, and dividing the header by a diaphragm; the water passes down the front portion of the header, flows through the inner tube, returns through the outer tube, and passes up through the back portion of the header. The advantage claimed for this boiler is that the tubes can quickly be removed for inspection and be replaced, while with most other water-tube boilers a tube can only be removed by cutting it out, and inserting a new one. The water in the Niclausse boiler cannot periodically be blown off, as the lower ends of the inclined tubes are closed; this may be considered a disadvantage. This boiler has had considerable success in France, but has made little headway in this country.

The **Belleville** boiler has no headers back or front. Each tube, which is of a zig-zag form, receives its water at the lower end, and delivers the steam and water in a state of froth or foam into a reservoir placed high up at the front end of the boiler, and at right angles to the tubes. This boiler requires very careful stoking.

BOILERS.

Stirling, Thornycroft, and **Yarrow** boilers are all variants of the same type of boiler. They have upper and lower drums connected by small tubes, either inclined or bent. These boilers have no headers such as are used in the Babcock and Niclausse boilers. The tubes of the Yarrow boiler are straight, and so give facilities for cleaning; those in the Thornycroft and Stirling boilers are bent. All these boilers have proved successful; the Thornycroft and Yarrow are used chiefly for marine work, where distilled water is used. There are many other water-tube boilers, such as the Hornsby, Climax, and others, but they vary chiefly in design, and not in principle.

The disadvantage of the earlier water-tube boilers was their lack of steam and water space, so that if an extra demand for steam was made upon them, there was a tendency to prime, or, in other words, for small particles of water to be carried off with the steam. They also required more attention from the stoker to keep the steam pressure and water level constant than a boiler of the Lancashire or Scotch marine type. In the Babcock boiler these disadvantages have been overcome by using a large steam drum; in the larger boilers of this make two drums are provided, placed side by side.

Strength of Boilers.—Some simple calculations as to the strength of boilers which a young engineer may be called upon to make will now be explained.

The force exerted by the steam acting at right angles to the surface of the water, and tending to burst or tear a boiler longitudinally, is found by multiplying the internal diameter of the boiler in inches by the steam pressure in pounds per square inch. To resist this bursting force there is the thickness of the metal plate on each side of the boiler, so that, ignoring for the moment the question of riveting, the stress per square inch in the metal plate forming the shell of the boiler can be found thus—

$$\frac{D \times P}{2 \times T} = \text{stress};$$

where D = diameter of the boiler in inches.
P = pressure of the steam in lbs.
T = thickness of the plate in inches, or parts of an inch.

Example.—What is the stress in the metal plates (undrilled portion) of a boiler 8 feet in diameter working at 150 lbs. pressure, the plates being ¾ inch, or ·75 inch thick? The calculation is

$$\frac{96 \times 150}{2 \times \cdot 75} = 9,600 \text{ lbs.},$$

or about 4·3 tons per square inch.

The calculation does not, however, take into account the fact that the plate is weakened by the holes which have been drilled in it for the riveting. In order to find the stress upon the weakest portion of the boiler-shell, another factor is introduced, and the formula reads—

$$\frac{D \times P}{2\,T \times K};$$

where K = the percentage of strength of the joint, the method of finding which will be given later.

Example.—What is the stress in the weakest part of the plate of a boiler 8 feet in diameter working at 150 lbs. pressure, the plates being ¾ inch thick, and the percentage of strength of the treble-riveted butt joint 79 per cent? The calculation is

$$\frac{96 \times 150}{2 \times \cdot75 \times \cdot79} = 12{,}152 \text{ lbs.,}$$

or about 5·4 tons per square inch.

It is considered that the tensile stress upon mild steel plates of Cornish or Lancashire boilers having butt joints at the longitudinal seams should not exceed 12,500 lbs. per square inch, so that the thickness of the boiler shell in the example just given is about right. The figure, 12,500, is arrived at thus:—The tensile strength of the boiler plate is probably about 28 tons, and, allowing a factor of safety of about 5, we get the figure in question.

To find the thickness of shell for any given steam pressure, assuming we decide to allow a stress of 12,500 lbs. per square inch, the formula would be transposed thus—

$$\frac{D \times P}{2 \times 12{,}500 \times K} = \text{thickness.}$$

Example.—How thick should be the shell of a boiler 8 feet in diameter for a working pressure of 200 lbs.?

$$\frac{96 \times 200}{2 \times 12{,}500 \times \cdot79} = \cdot97 \text{ thick,}$$

so that a plate 1 inch thick would be used. It must be noted that a strength of 79 per cent. is only obtained with a treble-riveted butt joint; with a double-riveted butt joint the strength would only be about ·75 that of the undrilled plate.

Apart from the effort of the steam to burst the boiler longitudinally, there is the effort to extend the boiler lengthwise, or to pull it into two parts, owing to the pressure of steam on the ends. The stress on the plates due to this effort is, however, only half the stress due to the pressure acting in the direction previously considered, so that if the plates are strong enough to withstand the radial pressure, they are of ample strength for the tension due to the pressure on the ends of the boiler.

Dealing now with the *strength of the boiler ends*, it may be said that the thickness of flat-stayed surfaces has been arrived at from the results of experience, and that empirical formulæ have been arranged to suit. The formula given by Mr. Hillier, chief engineer of the National Boiler Insurance Co., is simple and corresponds very nearly with actual practice; it is as follows:—

$$P = \frac{C \times T^2}{S};$$

where P = suitable working pressure in lbs. per square inch.
C = a constant given below.
T = thickness of plate in sixteenths of an inch.
S = area supported by one stay.

In Cornish or Lancashire boilers S = the area of the largest circle which can be got between the gusset stays, and C = 220 approximately.

Example.—In the Lancashire boiler illustrated, the end plates are 1¾ inch thick, and the largest circle which can be got between the gusset stays is 16 inches (area 201). What may the working pressure of the boiler be? The calculation is

$$\frac{220 \times 13^2}{201} \text{ or } \frac{220 \times 169}{201} = 185.$$

The answer is 185 lbs. per square inch.

If, instead of being flat, the ends are dished outwards, we can find the stress per square inch in the plate by the following formula:—

$$\frac{R \times P}{2 \times T} = \text{stress};$$

where R = radius in inches of the dished end.
P = pressure of the steam in lbs. per square inch.
T = thickness of the plate in inches.

It is not advisable to allow so great a tensile stress in plates which have been dished when red hot, as in plates which have merely been rolled to form the shell, and a stress of 6,000 to 7,000 lbs. per square inch is considered about right. If we decide to allow a stress of 6,000 lbs. per square inch, and wish to know the thickness of a dished end, the above formula is transposed thus—

$$\frac{R \times P}{6{,}000 \times 2} = T.$$

Example.—How thick should be the end plate of a boiler, 8 feet in diameter, if the plate is dished out at a radius of 4 feet, allowing a stress of 6,000 lbs., the boiler pressure being 185 lbs. ?

$$\frac{48 \times 185}{6,000 \times 2} = \cdot 74 \text{ thick.}$$

If the end is flatter and dished out to a radius of 6 feet, the calculation will be—

$$\frac{72 \times 185}{6,000 \times 2} = 1 \cdot 1 \text{ thick.}$$

If the end is still flatter and is dished out to a radius of 8 feet, viz., the same as the diameter of the boiler, the calculation will be—

$$\frac{96 \times 185}{6,000 \times 2} = 1 \cdot 48 \text{ thick.}$$

The reason for multiplying the pressure by the radius, when dealing with dished ends, instead of multiplying the pressure by the diameter, as when dealing with cylinders, is because a sphere is twice as strong as a cylinder, and the radius is, of course, only half the diameter of a circle.

Strength of the Flues.—The pressure of steam on the flue tends to make it collapse; if the flue were truly cylindrical, and the metal had no tendency to buckle, the formula $\frac{D \times P}{2 \times T}$ would hold good, the metal being in compression; but, as a fact, the flues have a tendency to buckle, and the right thicknesses and distance apart of the Adamson rings have been found by experience.

Fairbairn's rule, which is sometimes quoted, was based on experiments made many years ago with plain tubes, and the rule does not hold good for flues having Adamson rings.

The Board of Trade formula, which is somewhat elaborate, is given at the foot, but the following table will probably be of greater service to the beginner. The table gives the distance which the Adamson rings should be placed apart from one another for various steam pressures, and the stress per square inch which may be allowed upon the metal forming the flue.

Working pressure $= \frac{9,900 \times T}{3 D} \left(5 - \frac{L + 12}{60 T} \right)$;

where T = thickness of tube in inches.
 L = distance between flanges in inches.
 D = outside diameter of tube in inches.

Note.—L must not be greater than 120 T − 12.

Steam Pressure.	Distance Apart of Rings.	Stress per Square Inch on Metal.
Lbs.	Ft. Ins.	Lbs.
100	3 0	3,200 to 4,500
120	2 8	4,000 ,, 5,000
160	2 3	4,750 ,, 5,700
200	2 1	5,000 ,, 6,000

Strength of Riveted Joints.—The strength of a riveted joint, as compared with that of a solid plate, depends upon the amount of metal which is left after drilling the rivet holes, and upon the strength of the rivets to withstand shear. The strength of the drilled plate is found thus—

$$\frac{P - D}{P};$$

where P = pitch of the rivet.
D = diameter of the hole drilled for the rivet.

Example.—What is the strength of a ⅞-inch plate which has been drilled with 1-inch holes at 4-inch pitch for a double-riveted butt joint?

$$\frac{4 - 1}{4} = \cdot 75.$$

The answer is ·75, so that the plate is only ·75 times as strong as an undrilled plate. If we take the case of a plate which has been drilled for a single-riveted lap joint—viz., with 1-inch holes to 2¼-inch pitch—we shall see that it is still weaker, thus—

$$\frac{2 \cdot 25 - 1 \text{ inch}}{2 \cdot 25} = \cdot 555,$$

so that the plate is only ·555 times as strong as an undrilled plate.

The strength of rivets in ordinary practice is a little greater than that of the drilled plates. The strength of rivets in a lap joint is found by the following formula :—

$$\frac{A \times N}{P \times T} = \text{strength of rivets as compared with the drilled plate};$$

where A = area of rivet.
N = number of rows of rivets.
P = pitch in inches.
T = thickness of plate in inches.

Example.—What is the strength of the rivets in a ⅝-inch plate drilled for 1-inch rivets, 2¼-inch pitch?

$$\frac{\cdot 78 \times 1}{2 \cdot 25 \times \cdot 625} = \cdot 557.$$

The rivets are, therefore, ·557 times as strong as the undrilled plate.

In the case of butt joints with two straps, the rivets are in double shear, and should be twice as strong as those in single shear, but as the Board of Trade reckon that a rivet which has to be sheared through in two places is only 1·75 times as strong as a rivet in single shear, this figure is usually taken, and the formula for finding the strength of rivets in double shear is as follows:—

$$\frac{A \times N \times 1\cdot 75}{P \times T}.$$

Example.—What is the strength of the rivets in a ⅞-inch plate drilled with 1-inch holes at 4-inch pitch for a double-riveted butt joint? The calculation is

$$\frac{\cdot 785 \times 2 \times 1\cdot 75}{4 \times \cdot 875} = \cdot 785.$$

The rivets in this case are ·785 times the strength of the undrilled plate.

The size of rivets and the pitch usually adopted for various joints and for various thicknesses of plate are given in Chapter ii.

PROPERTIES OF STEAM.

Steam.—Before discussing the evaporative power of boilers, chimney draught, &c., a few words dealing with the raising of steam may usefully be said. A clear idea as to what is meant by the latent heat of steam is essential to anyone calling himself an engineer.

If heat is applied to a boiler which is open to atmospheric pressure, say 14·7 lbs. per square inch (the atmospheric pressure varies, and is shown by the height of the barometer), a thermometer placed in the water will show a rise of temperature corresponding with the amount of heat put in until steam begins to form. Under atmospheric pressure in normal conditions this will take place at a temperature of 212° F., but before 1 lb. of water can be evaporated at this pressure, a very large further instalment of heat will require to be put into the water; this heat will not be shown by the thermometer. The heat, however, is there, and is called the latent heat of steam.

It is necessary to know that the quantity of heat required to raise the temperature of 1 lb. of pure water at its greatest density by 1° F. is called a British thermal unit, or B.T.U. What this thermal unit is equivalent to will be explained later. Now, if there were no such thing as latent heat of steam, the number of thermal units required to raise 1 lb. of water from 32° to 212° and turn it into steam would be 180. As a fact, the total heat which is required to raise 1 lb. of water from 32° to 212°, and turn it into steam at 14·7 lbs. pressure, is 1,146·1 thermal units. The difference—viz., 966·1—between the two sets of figures is the latent heat in 1 lb. of steam at a pressure of 14·7 lbs. per square inch. The total heat of steam at 14·7 lbs. pressure is 1,146·1 B.T.U.

If, instead of taking the case of heat applied to water at atmospheric pressure, we imagine heat applied to water under a constant pressure of 120 lbs. per square inch absolute—by absolute pressure we mean a pressure reckoned from a perfect vacuum, and not above the atmospheric pressure—the thermometer will rise steadily until it shows a temperature of 341° F., at which temperature steam will begin to be formed; but to raise the temperature of 1 lb. of water from 32° F., and turn it into steam under a constant pressure of 120 lbs., 1,185·4 thermal units will be absorbed.

If water is under a constant pressure of 200 lbs., and heat is applied, the thermometer will rise to 381·7° before steam begins to be formed, and 1,197·8 thermal units will be required to turn 1 lb. into steam. If, on the other hand, water is placed under a vacuum and heat is applied, steam will begin to form at a temperature much below 212° F. If a closed glass vessel, containing water, is held in the hand, the upper portion of the vessel being placed in communication with a very effective air pump, and an extremely good vacuum is formed, the temperature of the hand is sufficient to cause the water to boil. The temperature at which water boils is, therefore, dependent on the pressure upon it.

Table IX. gives the temperature of steam corresponding with the pressure, also the total heat contained in 1 lb. of the steam. The steam pressures are given in two columns, one giving the gauge pressure, or pressure above the atmosphere; the other the absolute steam pressure.

Now, with regard to the British thermal unit, the energy required to raise a weight of 1 lb. to a height of 1 foot is 1 foot-lb.; the energy required to raise 1 lb. 10 feet, or 10 lbs. to a height of 1 foot, is 10 foot-lbs. Joule found that 772 foot-lbs. of work were the equivalent of one thermal unit. That is to say, if 772 foot-lbs. of work are expended on a lb. of water, by violently agitating it with paddles, or by other means, the temperature of the water is raised by 1° F. The work of 772* foot-lbs. is called the mechanical equivalent of heat.

If the amount of heat thus put into 1 lb. of water, could conveniently be utilised, it would be capable of lifting a weight of 1 lb. through a height of 772 feet (assuming no friction); after

* More recent experiments show that this figure is rather low, and that 778 foot-lbs. is a more accurate figure.

doing this work the water would be back again at its original temperature.

TABLE IX.—PROPERTIES OF SATURATED STEAM.

Steam Pressure above Atmosphere, taking Atmosphere at 15 lbs.	Steam Pressure Absolute.	Temperature.	Total Heat in 1 lb. of Steam from 32° F.	Volume of 1 lb.	Steam Pressure above Atmosphere, taking Atmosphere at 15 lbs.	Steam Pressure Absolute.	Temperature.	Total Heat in 1 lb. of Steam from 32° F.	Volume of 1 lb.
Lbs. per Sq. In.	Lbs. per Sq. In.	Degs. F.	B.T.U.	Cub. Ft.	Lbs. per Sq. In.	Lbs. per Sq. In.	Degs. F.	B.T.U.	Cub. Ft.
	1	102·1	1112·5	330·36	85	100	327·9	1181·4	4·33
	2	126·3	1119·7	172·08	90	105	333·1	1182·4	4·14
	3	141·6	1124·6	117·52	95	110	334·6	1183·5	3·97
	4	153·1	1128·1	89·62	100	115	338·0	1184·4	3·80
	5	162·3	1130·9	72·66	105	120	341·1	1185·4	3·65
	6	170·2	1133·3	61·21	110	125	344·2	1186·4	3·51
	7	176·9	1135·3	52·94	115	130	347·2	1187·5	3·38
	8	182·9	1137·2	46·69	120	135	350·1	1188·2	3·27
	9	188·3	1138·8	41·79	125	140	352·9	1189·0	3·16
	10	193·3	1140·3	37·84	130	145	355·6	1189·9	3·06
	11	197·8	1141·7	34·62	135	150	358·3	1190·7	2·96
	12	202·2	1143·0	31·88	140	155	361·0	1191·5	2·87
	13	205·9	1144·2	29·57	145	160	364·3	1192·2	2·79
	14	209·6	1145·3	27·61	150	165	366·0	1192·9	2·71
	14·7	212·0	1146·1	26·36	155	170	368·2	1193·7	2·63
	15	213·1	1146·4	25·85	160	175	370·8	1194·4	2·56
1	16	216·3	1147·4	24·24	165	180	372·0	1195·1	2·49
2	17	219·6	1148·3	22·89	170	185	375·3	1195·8	2·43
3	18	222·4	1149·2	21·70	175	190	377·5	1196·5	2·37
4	19	225·3	1150·1	20·64	180	195	379·7	1197·2	2·31
5	20	228·0	1150·9	19·72	185	200	381·7	1197·8	2·26
10	25	240·1	1154·6	15·99	190	205	383·8	1198·4	2·21
15	30	250·4	1157·8	13·46	195	210	385·8	1199·1	2·16
20	35	259·3	1160·5	11·65	200	215	387·8	1200·1	2·12
25	40	267·3	1162·9	10·27	205	220	389·9	1200·3	2·08
30	45	274·4	1165·1	9·18	235	250	401·1	1203·7	1·82
35	50	281·0	1167·1	8·31	285	300	417·5	1208·7	1·53
40	55	287·1	1169·0	7·61	335	350	430·1	1212·6	1·32
45	60	292·7	1170·7	7·01	385	400	449·9	1217·1	1·16
50	65	298·0	1172·3	6·49	435	450	456·7	1220·7	1·05
55	70	302·9	1173·8	6·07	485	500	467·6	1224·0	·94
60	75	307·5	1175·2	5·68	585	600	487·0	1229·9	·80
65	80	312·0	1176·5	5·35	685	700	504·1	1235·1	·68
70	85	316·1	1177·9	5·05	785	800	519·5	1239·8	·60
75	90	320·0	1179·1	4·79	885	900	533·6	1244·2	·54
80	95	324·1	1180·3	4·55	985	1000	546·5	1248·1	·49

Steam in contact with water is called saturated steam; this is not the same thing as the steam being wet. Wet steam is steam in which particles of water have become entrained through the steam leaving the water at too great a velocity, or through the temperature having become lowered, and some of the steam having turned back into water. Dry saturated steam is steam which is free from particles of water. Superheated steam is steam to which further heat has been added after it has been formed. Such steam, if placed in contact with water, will absorb a part of it, turning the water into steam; the temperature of the steam falls slightly during the process. This addition of heat to the steam, or superheating it, after it has been formed, does not increase its pressure (apart from the pressure due to increase of volume), and unlike latent heat, it is shown by a thermometer.

Evaporation of a Boiler.—The evaporation of a boiler as given by the makers is usually in terms of pounds of water evaporated "from and at 212° F." This means that so many pounds of water will be evaporated, if it is fed into the boiler at a temperature of 212°, and is evaporated at atmospheric pressure. If, however, the water is fed into the boiler at a lower temperature, and is evaporated under a considerable pressure, the evaporation of the boiler will be less than that given by the maker. Thus, if water is fed into the boiler at a temperature of 60° F., and is evaporated under a pressure of 120 lbs., the amount of water evaporated will be only $\frac{10}{12}$ that given by the makers. If the water is fed in at 150° F., and evaporated at 120 lbs., the evaporation will be about $\frac{11}{12}$. If fed in at 32° F., and evaporated at 200 lbs. pressure, the evaporation will be $\frac{10 \cdot 0}{12 \cdot 4}$ of that given.

The formula to enable one to find out what the actual evaporation will be, if the evaporation "from and at" is known, is as follows:—

$$A = \frac{B + 32 - C}{966};$$

where A = the factor by which the lbs. given as "from and at" are divided.
B = the total heat of the steam at the working pressure.
C = the temperature of the feed water.

Example.—If the makers say that a boiler will evaporate 1,200 lbs. of water per hour from and at 212°, how many lbs. will it evaporate if the working pressure is to be 120 lbs. (or 135 lbs. absolute) and the temperature of the feed is only 60° F.? The reader will see from Table IX. that the total heat of steam at 120 lbs. pressure (above atmosphere) is 1,188 B.T.U. The formula then, in actual figures, is

$$A = \frac{1{,}188 + 32 - 60}{966} = 1\cdot 2.$$

The answer is 1·2, so that, if we divide the 1,200 lbs. given by the makers by 1·2, we get 1,000 lbs., which is the amount of water the boiler will evaporate under the conditions given.

If the actual evaporation is known, and it is desired to ascertain what the equivalent evaporation "from and at 212°" would be, the number of lbs. actually evaporated is multiplied by the factor obtained by the formula given above.

Coal and its Evaporative Power.—We have seen what a certain number of B.T.U. are capable of doing in the way of raising steam; this information, however, is of but little practical use to us, unless we know how many B.T.U. we can get from a pound of coal. A pound of coal burnt in the laboratory gives from 13,000 to 15,000 B.T.U. A pound of pure carbon gives 14,500 B.T.U., but coal contains other constituents than carbon, such as hydrogen, oxygen, nitrogen, sulphur, and ash. The B.T.U. values of 1 lb. of some of the best known coals, and of petroleum, are approximately as follows:—

Powell Duffryn,	15,500
Nixon's Navigation,	15,000
Newcastle,	14,820
Derbyshire, Yorkshire,	13,860
Scotch,	13,500
Coke,	12,820
Petroleum,	17,800 / 20,000

Steam coals are those which contain a large proportion of fixed carbon and a small proportion of volatile constituents. They burn without giving off much gas. A coal which fulfils these conditions to the fullest extent is anthracite; the largest proportion of coal is, however, bituminous or smoky coal.

Powell Duffryn is almost, but not quite, pure anthracite; it contains about 88·24 per cent. of carbon. Nixon's Navigation is semi-anthracite. The others given above are bituminous; they contain from 75 to 83 per cent. of carbon, the remainder consisting of various volatile constituents and ash. Coke contains from 86 to 88 per cent. of carbon.

Coals are sometimes divided into caking and non-caking coals; the former soften when heated, and form a spongy mass; in non-caking coals the particles remain separate and allow the air to pass between them. Cannel coal, which is very rich in volatile constituents, is used for producing gas for lighting purposes.

The ash in Powell Duffryn coal is about 3·26 per cent. In Nixon's Navigation it is about 7 per cent., while in inferior coals it may amount to 10 per cent. or more.

One lb. of coal, yielding 14,007 B.T.U., is theoretically capable of evaporating 14·5 lbs. of water from and at 212° F. In actual practice 1 lb. of coal evaporates between 8 and 10 lbs. of water. Why, it may be asked, this great difference? In the first place there is the correction to be made for the fact that the feed may not be so hot as 212°, and for the fact that the water will, of course, be evaporated at a much higher pressure than that corresponding with 212°. The formula for making the necessary correction has already been given. Then there is the loss due to the heat which passes away in the gases to the chimney. The temperature of these gases must not be lower than that of the steam in the boiler, otherwise they would be harmful. The losses from this source are probably from 12 to 15 per cent., and may easily be greater. The other losses are those due to incomplete combustion, radiation, &c.

An economiser, which will be described later, will extract and make use of some of the heat of the waste gases, and if the boiler itself extracts 70 per cent. of the heat theoretically contained in the coal it may be considered efficient.

When boilermakers say that a boiler will evaporate so many lbs. of water per lb. of combustible, they mean that the weight of the unconsumable ash must be deducted from the weight of the coal burnt.

Rate of Combustion.—The number of pounds of coal which can be burnt on each square foot of grate area depends on the kind of coal and upon the draught. The draught given by a chimney depends upon its height, assuming that its area is sufficiently large to carry off the gases. Draught caused by the height of the chimney alone is called natural draught[*]; if a fan is put inside

[*] The formula to enable the actual draught to be ascertained is as follows:—

$$D = H \left(\frac{7·6}{A + 461}\right) - \left(\frac{7·9}{B + 461}\right);$$

where H = height of chimney in feet.
A = temperature of external air.
B = temperature of gases in chimney.
D = draught.

Example.—What will be the draught in a chimney, 100 feet high, when the temperature outside is 60°, and inside 400°?

$$100 \left(\frac{7·6}{60 + 461}\right) - \left(\frac{7·9}{400 + 461}\right) = ·542.$$

The answer is ·542 inch.

the flue to increase the draught, the latter is called induced draught; if a fan is employed to increase the pressure of air in the stokehold of the furnace the draught is called forced draught.

The intensity of the draught is spoken of as so many inches, or parts of an inch, of water. This means that if a U-shaped tube is partly filled with water, and one leg of the U is placed in communication with the chimney, the other leg being open to the atmosphere, the water in the leg connected to the chimney rises above the level of the water in the other leg. If the difference is ¼ inch the draught is said to be equal to ·25 inch of water.

With a properly constructed chimney,* 100 feet high, and with gases at a temperature of 400° above that of the atmosphere outside, say 60° F., a draught of about ·54 of an inch is obtained. When fans are employed to give forced draught the draught is usually between 1 and 4 inches. If the draught is forced to too great an extent it is injurious to the boiler, owing to the intense heat caused.

We have said that the amount of coal that can be burnt on each square foot of grate area depends on the kind of coal and on the draught. The following are approximately the amounts of coal which can be burnt per square foot of grate, with a draught of ·5 inch, and with hand firing.

Coke,	11–12 lbs. per hour.
Nixon's Navigation,	18–20 ,, ,,
Powell Duffryn,	20–23 ,, ,,
Bituminous coal,	20–32 ,, ,,

The nearer the coal approaches the qualities of anthracite the greater is the draught required. Bituminous coal requires less

---

* A good empirical formula for determining the most suitable area for a chimney is the following:—

$$A = \frac{W}{16 \sqrt{H}};$$

where A = area in square feet.
W = weight of coal burnt per hour.
H = height of chimney in feet.

Example.—What must be the area of a chimney, 150 feet high, for boilers burning 2,000 lbs. of coal per hour?

$$A = \frac{2,000}{16 \sqrt{150}} \text{ or } A = \frac{2,000}{16 \times 12 \cdot 24} = 10 \cdot 2 \text{ feet.}$$

The answer is 10·2 square feet.

A rule often worked to is 6 square feet of chimney area for every 30 feet × 8 feet Lancashire boiler up to four boilers; beyond this 5 square feet for every additional boiler.

draught; with a draught of ·25 inch, from 16 to 20 lbs. of bituminous coal can be burnt per square foot of grate area per hour.

The ratio of grate area to heating surface varies, or should vary, according to the kind of fuel to be consumed and the draught. The ordinary ratios in Lancashire, water-tube, and locomotive boilers are approximately as follows:—

	Grate Area.	Heating Surface.
Locomotive boiler,	1	23–28
Water-tube boiler (Babcock type),	1	50–70
Locomotive boiler,	1	60–80

The evaporation per square foot of heating surface varies considerably in different types of boiler. Makers of Lancashire and Cornish boilers usually allow about 1 square foot of heating surface for every 5 to 10 lbs. of water to be evaporated from and at 212° F. Makers of the Babcock boiler allow about 1 square foot for every $2\frac{1}{2}$ to $3\frac{1}{2}$ lbs. to be evaporated. In locomotive boilers with a strong draught 1 square foot of heating surface will evaporate from 10 to 17 lbs. of water. The actual evaporation depends on the amount of coal properly burned in relation to the heating surface, and upon the degree of effectiveness of the heating surface.

In order that coal may be burnt to the best advantage, the right amount of air must be supplied, and there must be sufficient space in which combustion can be carried out. If insufficient air is supplied, or if combustion is interfered with by contact with a cool surface, incomplete combusion takes place, and carbon monoxide or CO is formed. Now, if 1 lb. of carbon is improperly burned so that it forms CO, it will give only about 4,330 B.T.U., as against 14,500 B.T.U., which it gives if properly burnt so as to form carbon dioxide or CO_2. If, on the other hand, too much air is admitted to the furnace, the temperature is reduced, and some of the heat given out by the coal is wasted. Smoke and soot are produced by the incomplete combustion of the carbon.

The number of heat units which can be transmitted through a steel boiler plate depends principally upon the difference in temperature of the gases on one side of the boiler plate and of the water on the other side, and in a less degree upon the thickness of the plate.

Mr. Blechynden, in a paper read before the Institute of Naval Architects some years ago, gave the results of a series of exhaustive experiments which he made as to the transmission of heat through plates of various thicknesses and with different degrees of temperature on the two sides of the plate. The following are some of the results obtained :—

Difference of Temperature of the Two Sides of Plate.	Heat transmitted per 1° difference per square foot per hour.	
° F.		
848	12·78	Plate 1·1875 inch thick.
1,013	15·26	,, ,,
1,278	20·9	,, ,,
626	10·89	Plate ·75 inch thick.
1,058	19·18	,, ,,
1,233	21·92	,, ,,
563	11·90	Plate ·5625 inch thick.
1,148	25·7	,, ,,
503	11·81	Plate ·25 inch thick.
723	16·55	,, ,,
893	20·65	,, ,,
738	16·46	Plate ·125 inch thick.
1,083	25·48	,, ,,

In a paper read before the same society a year previously, Mr. (now Sir John) Durston gave the results of some experiments undertaken to show the effect of grease on the surface of boiler plates; the experiments showed that a film of grease caused a most astonishing decrease in the number of heat units which could be transmitted through a plate of a given thickness, and with a given difference of temperature on its two sides.

Feed Water for Boilers and Boiler Compositions.—It is important that the feed water for boilers should be as free from impurities as possible. Water usually contains a certain amount of lime, magnesia, and other impurities; these are thrown down by the heat in the form of sediment, which becomes extremely hard, and has an injurious effect on the boiler. The incrustation acts as a non-conductor of heat, so that the water is unable to take away the heat from the steel boiler plates, and the metal may in consequence get unduly hot; this in turn may cause the furnace crown to collapse, or may give rise to undue expansion, and thus set up dangerous stresses in the boiler. The non-conducting properties of this scale may also cause a serious falling off in the evaporative power of the boiler.

It is considered that an incrustation $\frac{1}{8}$ inch thick causes a falling off of 15 per cent. in the evaporation, and consequently 15 per cent. of the coal is wasted.

As previously mentioned, the impurities in water are thrown down by heat. This fact is made use of in the Niclausse boiler, in which the feed water is admitted to the steam space in the form of spray over a suitable tray provided with a sludge pipe. The impurities fall into the tray in the form of sludge, and the latter is, or should be, periodically blown out. The Boby feed-water heater and detartariser acts on the same principle. Various mechanical filters are made for removing mud and impurities merely held in suspension. In these provision is made for cleaning the filtering material, either by blowing through steam or by reversing the direction of the current of water.

Various boiler compositions to prevent incrustation are sold; the effect of these is to prevent the sediment forming a hard scale, but these compositions should not be used unless the water has been analysed, and the user has reasonable grounds for believing that the composition will do what is claimed for it. Most of the compositions contain tannic acid. This acid is effective in cases where the water contains carbonate of lime and magnesia, but if used in excess is injurious to the boiler plates. Soda is useful when the water contains sulphate of lime or acids, but it causes a boiler to prime if used in excess. The right course, if the water is hard or impure, is to treat it chemically, or by filtration before it enters the boiler.

Testing Boilers.—Before a boiler leaves the makers' works it is tested by hydraulic pressure for tightness and strength. The pressure usually employed is 50 per cent. greater (sometimes 100 per cent. greater) than the working pressure for which the boiler has been constructed. After a boiler has been supplied and fixed, the purchaser may desire to test it for its evaporative performance. This is done by weighing or measuring the water pumped in, the coal consumed, and the weight of the ashes removed. Care is taken to see that the level of the water is the same at the end as at the beginning of the trial, and that the thickness of the fire is approximately the same at the commencement and end of the trial.

The results of such a trial are usually tabulated in the manner shown by Table x. Trials are often carried out in a more elaborate manner than is indicated by the table. For instance, the coal may be analysed to ascertain the number of B.T.U. yielded by it. The waste gases may also be analysed, and the steam be tested for dryness. Steam is tested for dryness usually by means of a throttle calorimeter. The principle of the calorimeter is this:—When steam passes through a constricted opening, it is said to be wire-drawn, and becomes slightly superheated, the amount of the superheat depending on the dryness of the steam. Now, the amount by which really dry steam becomes superheated being known, if the amount of superheat imparted to the steam under test is ascertained, it is not difficult to calculate the percentage of moisture in it.

Trial of a Water-tube Boiler for a 250 I.H.P. Compound Non-condensing Engine.

Date of test,	22/8/06.
Duration of test,	8 hours.
Heating surface,	1,506 square feet.
Grate area,	28 square feet.
Ratio of heating surface to grate area,	53·8 to 1.
Average gauge pressure,	180 lbs.
Temperature of feed water,	158° F.
Pounds of coal burnt,	4,190·4.
,, refuse,	297·5.
,, combustible,	3,892·9.
Per cent. of ashes,	7·1.
Coal burnt per square foot of grate,	18·7 lbs.
Total water evaporated,	38,552 lbs.
Water evaporated per hour,	4,819 lbs.
Water evaporated per pound of coal under actual conditions,	9·2 lbs.
Water evaporated per pound of coal from and at 212° F.,	10·3 lbs.
Water evaporated per pound of combustible from and at 212° F.,	11·1 lbs.
Temperature of flue gases,	450°.
Draught in inches of water,	·5 inch.
Efficiency of boiler, assuming a basis of 14,000 B.T.U. per pound of coal,	71 per cent.

Note.—If an economiser is fixed in connection with the boiler, its heating surface is given, and the temperature of the water and gases before entering and after leaving are also given. If a superheater is used, its heating surface is given, and the temperature of the steam as it leaves the boiler is also noted.

Results similar to those given in the table would be expected from a good Lancashire boiler, but the heating surface would be less, and the temperature of the flue gases would probably be higher.

CHAPTER IV.

STEAM-RAISING ACCESSORIES.

We will now consider some accessories which are used in connection with steam boilers.

Pumps.—In order to force water into a boiler under steam pressure, a pump or injector is required. When the boiler is used for driving a slow-speed engine, the latter is usually provided with its own pump, driven from an eccentric fitted on the shaft, or from some reciprocating part of the engine, so that at every stroke of the engine a small quantity of water is forced into the boiler. A by-pass arrangement is provided, so that any excess water may be returned to the hot well. When the engine

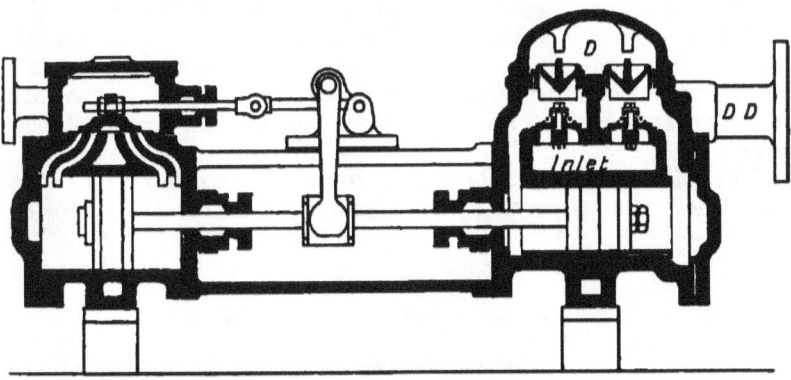

Fig. 21.—Worthington Duplex feed pump.

is of the high-speed type, or when the installation consists of several engines or turbines, a separate pump is usually employed. This pump may be driven by steam or by an electric motor; by running it faster or slower a greater or smaller quantity of water can be fed in to the boilers, and thus the water-level be kept constant.

When a steam pump is employed, a very common form is that known as the Duplex direct-acting pump, shown by Fig. 21. In this pump there are two steam and two water cylinders placed horizontally side by side; the steam piston is connected by a rod to the water piston or plunger. The area of the steam piston is usually twice that of the water piston, consequently the former is able to drive the latter. The motion of the rod of one steam cylinder gives the necessary motion to the valve of the other cylinder. The water enters through the space marked "inlet," is forced through the upper valves into the space marked D, and from thence it descends through a passage placed between the two pumps to the common discharge, D D.

The Duplex pump is inexpensive and reliable, but is wasteful of steam, as the steam is admitted right to the end of every stroke, and the steam clearance spaces are large.

A much more economical form of direct-acting steam pump is the Weir (Fig. 21a). In this pump there is usually one steam cylinder and one water cylinder arranged singly, as shown by the illustration, or in pairs. The valve is arranged to cut off steam when the piston has travelled about 75 per cent. of its stroke, so that during the last 25 per cent. of the stroke the steam is used expansively. Provision is made for admitting steam by by-passes during the whole of the stroke when the pump is first started; when the pump is well under weigh the by-pass valves are closed. There are two steam valves, one main and one auxiliary valve; the latter is actuated by a tappet motion from the pump rod, and admits steam to either side of the main valve; the latter is alternately driven backwards and forwards by the steam, and so controls the action of the pump.

The following is a table giving some of the standard sizes of the Weir pumps, having one steam cylinder and one water barrel :—

TABLE X.

Diameter of Steam Cylinder.	Diameter of Pump.	Length of Stroke.	Strokes per Minute.	Size of Suction.	Size of Delivery.	Gallons per Hour.
Inches.	Inches.	Inches.		Inches.	Inches.	
5	3½	6	44	1¼	1	515
6	4	8	40	1½	1½	810
6½	4½	10	36	2	1½	1,160
7	5	12	30	2½	2	1,450
8	5½	12	30	2½	2	1,730
8½	6	12	28	3	2½	1,950
9½	7	18	26	3½	3	3,700
10¾	8	18	24	4	3½	4,450
12	9	24	24	4½	3½	7,540

STEAM-RAISING ACCESSORIES.

If the reader checks the number of gallons delivered according to the table by the formula given later, he will find that a somewhat small allowance has been made for slip. If, however, the estimated slip should be exceeded, the number of gallons given

Fig. 21a.—Weir feed pump.

above can be obtained by running the pumps a little faster, as the piston speed indicated by the figures is very moderate indeed. The Weir pump is also made in the compound form —*i.e.*, having one H.P. cylinder, one L.P. cylinder, and two water cylinders.

Another very reliable form of direct-acting pump is the "Deane," made by the Pulsometer Engineering Company, but in this pump the steam is not used expansively.

Still another form of steam pump is that of the flywheel type, as shown by Fig. 22, and made by Messrs. Cameron, of Leeds, and others. In this pump there are one or more steam cylinders with corresponding water cylinders. The valve controlling the supply of steam to each cylinder is driven from an eccentric mounted on the shaft carrying the flywheel. With this form of pump the admission of steam need not be continued until the end of the stroke, so that it may be used expansively; further, compound cylinders may be used. The objection to the flywheel form of pump is that, should any temporary obstruction occur in the delivery pipes, the energy stored in the flywheel may cause an undue rise of pressure in the pipes.

In electric generating stations electrically-driven feed pumps are sometimes employed; a small electric motor running at a high speed drives the pump through suitable gearing at a moderate speed. There are, however, certain electrical difficulties in regulating the speed of the motor through a sufficiently wide range of speed, and a by-pass on the delivery side of the pump is a wasteful arrangement.

The number of gallons of water which should be discharged by a double-acting pump, if there were no slip, can be found thus—

Fig. 22.—Cameron pump.

STEAM-RAISING ACCESSORIES.

$$\frac{A \times L \times N}{277} = G;$$

where A = area of pump bucket or plunger in inches.
L = length of stroke in inches.
N = number of effective strokes per minute.
G = gallons per minute.

To obtain the number of gallons delivered *per hour*, the result must be multiplied by 60.

If the pump is of the ram type which draws in water once during every two strokes, the number of effective strokes is reduced by half. If the pump has more than one barrel, the result obtained by the above formula must be multiplied by the number of barrels. If it is desired to find the number of *pounds* of water discharged per minute, the number of gallons must be multiplied by 10.

The formula does not take into account slip—*i.e.*, the percentage of water which finds its way back while the valves are closing, or which leaks past the bucket or ram. It is, therefore, necessary to make an allowance for this, and in the case of the average boiler feed pump it is best to assume that only 90 per cent. of the amount which should theoretically be delivered is actually pumped. It should be stated that there are 277·27 cubic inches in a gallon, while in the formula the figure is given as 277. The latter figure is given for convenience of calculation; it is sufficiently accurate, considering that a margin must, in any case, be allowed for slip.

Speeds up to 100 feet per minute are suitable for the buckets or plungers of boiler feed pumps.

Injectors.—In locomotives the water is usually supplied by means of an injector. This instrument was invented by a French engineer named Giffard; by its use a boiler can feed itself with water without the intervention of a force pump.

An injector is shown diagrammatically in Fig. 23. The principle upon which it works is this:—The steam nozzle is surrounded by cool water at A; when the steam passing through the pipe B meets this water it condenses and forms a partial vacuum; this causes the steam in the pipe B to travel at a great velocity, in order to fill up the vacuum as the steam is at the full boiler pressure; the vacuum also causes the water in pipe C to rush in at a great velocity, the water being driven in by the pressure of the atmosphere. The velocity acquired by the column of steam and water is so great that it causes the pressure on the under side of the check valve to be higher than the pressure above it, the check valve opens, the condensed steam and water enter the boiler, and so keep it supplied.

A modern injector (made by Messrs. Holden & Brooke, of Manchester) is shown by Fig. 24. The injector is of the one-movement type—that is to say, the admission of steam and water is simultaneously regulated by the one handle.

On first starting an injector, before the column of condensed steam and water has acquired enough energy to open the check valve, it passes away through the overflow outlet, but as soon as the injector gets to work the overflow ceases.

An injector which is required to lift its own water, and to work with fairly high steam pressures, will not work satisfactorily if the temperature of the water at A is higher than about 80° F. If, however, the injector is arranged so that the water flows into it by gravity, injection water at a higher temperature may be used, especially if the working steam pressure is low. With a boiler pressure of 50 lbs. per square inch, the temperature of the injection water may be as high as 140° F. With a boiler pressure of 100 lbs., water at a temperature of 120° F. may be used. With a boiler pressure of 200 lbs., the temperature of the injection water should not be more than 90°, even if fed into the injector by gravity.

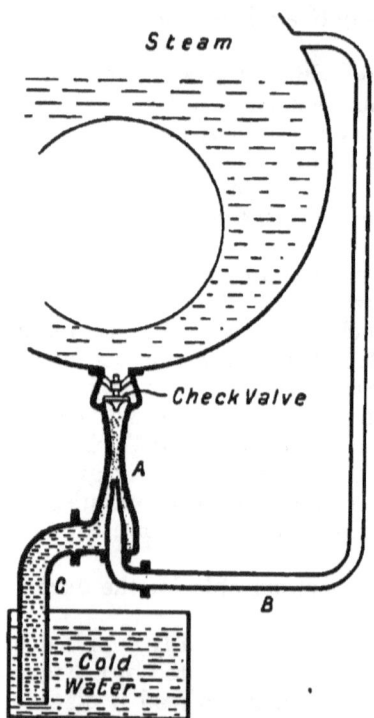

Fig. 23.—Diagram showing action of injector.

Exhaust Steam Injectors.—An injector may be worked by the exhaust steam from a non-condensing engine; such an arrangement has the effect of slightly reducing the back pressure in the engine and of warming the water. If, however, the load on the engine is very variable the action of the injector is rather uncertain. The objection to the use of injectors is that there is a danger of the passages becoming choked with sediment, in which case the injector naturally ceases to work.

Feed-Water Heaters and Economisers.—It is uneconomical to supply a boiler with cold feed water, and any heat which would otherwise be wasted should be used to heat it. When a

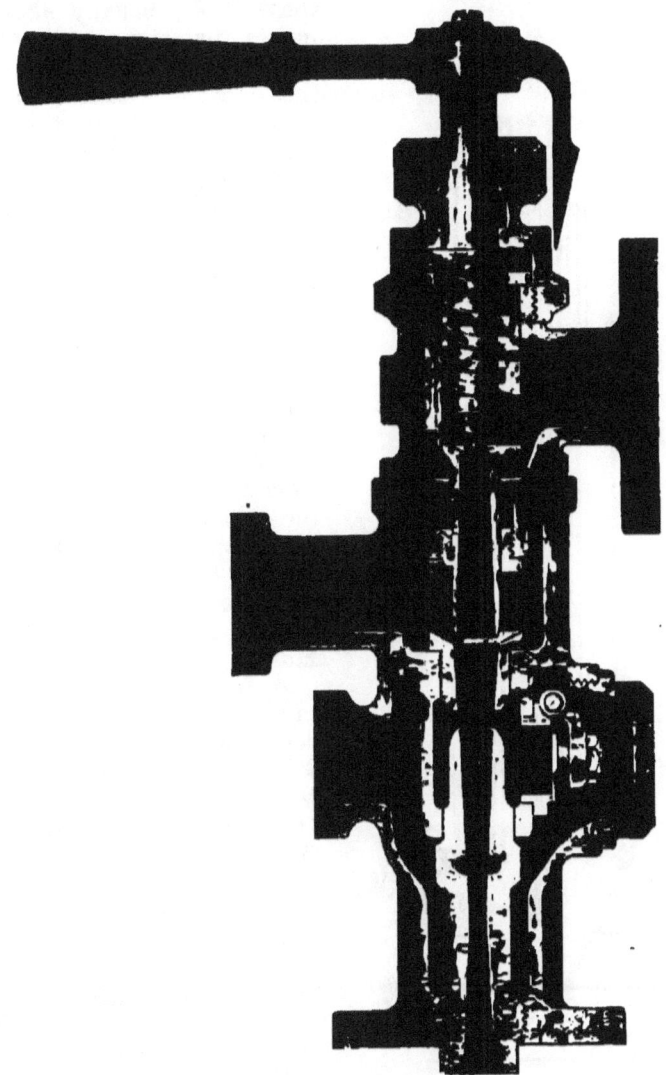

Fig. 24.—Modern injector.

boiler is used for driving a non-condensing engine, the steam, after it has done its work in the engine, can be utilised for warming the feed water by means of a simple apparatus called a feed-water heater. This consists of a number of tubes placed inside a shell or drum; through these tubes, usually about 1½ inches diameter, the water is forced on its way to the boiler, the exhaust steam from the engine is taken to the shell before passing away to the atmosphere; it surrounds the small tubes through which the water is passing, and gives up a large portion of its heat to the water.

A good feed-water heater is shown by Fig. 25. It will be seen that the tubes are free to expand irrespective of the outer casing. Any impurities which may be thrown down from the water by the heat of the exhaust steam will fall into the deposit chamber, from whence the deposit can easily be removed. The feed heater shown is made by Messrs. Holden & Brooke.

A feed-water heater of the kind shown will raise the temperature of the feed from about 60° F. to about 180° F., when worked at its rated capacity. If worked below its full capacity, the water can be made still hotter. As will be seen from the illustration the feed water passes once up and once down the tubes; in the latest heater made by Messrs. Holden & Brooke and called their High Velocity feed heater, the water is passed six times up and six times down the tubes; the water is also driven through the tubes at a high velocity, the result being that the water rolling over on itself, so to speak,

Fig. 25.—Feed-water heater.

in its passage through the tubes, extracts more heat from the steam. With this form of heater the temperature of the feed is raised from 60° to about 200° when worked at its full rated capacity, but more energy is expended in driving the water through. The saving effected through heating the feed water is approximately 1 per cent. of coal for every 10° by which the temperature of the feed is raised.

In cases where the exhaust steam is not available for this purpose, as in a condensing engine, the feed water is often heated by means of an **Economiser**. An economiser consists of a large number of vertical tubes about 4 inches in internal diameter and 9 feet long, placed in a brickwork setting at the back end of a Cornish, Lancashire, or other boiler; the hot gases are compelled to pass around these economiser tubes on their way to the chimney; the feed water is forced through the tubes before entering the boiler. It has been found necessary to provide means to keep the tubes free from soot; to effect this each tube is provided with a scraper which travels slowly up and down its exterior; the power to work these scrapers is taken from the engine by means of shafting, or a small electric motor is provided for the purpose.

If the gases leave the boiler at a temperature of 650°, an economiser will extract about 300°, reducing the temperature of the gases to 350°. The temperature of the feed water passing through the economiser will be raised by about 150°, say from 62° to 212°, or if the feed water is taken from the hot well, and its temperature is already 100°, the economiser will raise the temperature to 250°.

Economisers are not used so frequently with water-tube boilers as with those of the shell type, as the larger amount of heating surface in a boiler of, say, the Babcock type, extracts a larger percentage of the heat from the gases; in fact, the portions of the tubes in a Babcock boiler nearest the chimney may be looked upon in the light of an economiser, but without the mechanical scrapers. It is not advisable to reduce the temperature of the chimney gases below 350°, if a good draught is desired.

Thermal Storage.—Another method of heating the feed water is by means of live steam; this method is adopted in what is known as the Thermal Storage System, introduced by Mr. Druitt Halpin. The thermal storage system is frequently used in electric generating stations where, during the day time, the boilers can generate more steam than the engines or turbines require, but where, during the evening, when the heavy load comes on, the boilers have difficulty in meeting the demand for steam. The system briefly is this:—There is a large drum containing a considerable amount of feed water, the drum being cons ructed to bear the same pressure as the boiler. During the hours when the engines require only a small quantity of steam, a portion of the steam generated is passed into the thermal storage drum, and its heat is given up to the water. The amount of steam passed is such that by the time the evening comes on with a heavy demand for steam, the water in the drum will have received sufficient heat to enable it to turn into steam without the addition of any further heat, should the pressure fall; or the addition of a very small quantity of heat, obtained by passing the water from the storage drum through the boiler, will cause the water to evaporate at the full steam pressure.

Superheaters.—The advantage of superheating steam, from the point of view of economy, has long been known, but it is only within recent years that superheaters have come into anything like general use. By superheating steam before it leaves the boiler, loss due to condensation of the steam in the pipes conveying it to the engine is avoided, and initial condensation in the engine itself is reduced. As the initial condensation in a simple steam engine may be as much as 30 or 40 per cent., the gain through using superheated steam with such engines is apparent. In compound or triple-expansion engines, or in turbines, where the initial condensation is less, the gain through the use of superheated steam is not so great, but it is still considerable.

In trials of some high-speed triple-expansion engines, which were carried out a few years ago by the makers, the saving due to superheating the steam by 100° at the engine stop valve varied from $14\frac{1}{2}$ to $16\frac{1}{2}$ per cent. at full load, and from 20 to $22\frac{1}{2}$ per cent. at one-third load. Speaking generally, it may be said that with a good compound or triple-expansion engine a saving of 1 per cent. of steam at full load is effected by every 6 per cent. of superheat given to the steam, while in a steam turbine a saving of 1 per cent. is effected by every 10° to 12° of superheat.

The Admiralty recently carried out some trials to ascertain the value of superheated steam in H.M.S. *Britannia*, and it was found that at cruising speed, when the engines developed one-fifth of their power, the saving effected through superheating the steam by 83° at the engines was 15 per cent.

A superheater consists of a number of small tubes, which are placed in the path of the furnace gases, and through which the steam is passed. In Cornish and Lancashire boilers the superheater is usually placed in the flue at the end of the boiler; dampers are provided to prevent the steam becoming too highly superheated. If the steam is too highly superheated, it carbonises the oil used in the engine cylinder and valve chests, as the best mineral oil will not stand a temperature higher than 650°, and few oils will bear a temperature higher than 600°.

For reciprocating engines the amount of superheat usually employed is about 100°, and occasionally 150°; thus with steam at a pressure of 180 lbs. the total temperature is about 480°, and occasionally 530°. In a steam turbine the degree of superheat is limited by the temperature which the bronze blades will safely bear.

In the Babcock boiler the superheater, as will be seen by Fig. 20, is above the main steam-raising tubes and below the drum; in this position it is out of the way of the fierce heat of the furnace, and special dampers are not required. The small pipe leading from the drum to the superheater is for flooding the latter while steam is being raised.

In the M'Phail & Simpson superheater, which was probably the first really successful superheater used in this country, the steam, after being superheated in pipes placed in the flues of the boiler, was taken through a pipe which passed through the water space of the boiler; the effect of this was to take some of the sting out of the steam, or, in other words, to prevent an excessive degree of superheat being reached.

In the Cruse superheater the tubes are 6 inches diameter, and have an internal pipe made of copper through which a stream of water is passed. This water may be drawn from the water space of the boiler, from the economiser, or from the cold water feed mains, and in this way the amount of superheat can be regulated.

Some superheaters are separately fired. The best known superheater of this type is, perhaps, the Schmidt, in which there are two sets of coils, an upper and lower. The saturated steam enters at the top—*i.e.*, farthest from the fire—and travels down through the coils; dampers are provided, by means of which the degree of superheat can be regulated. A separately-fired superheater, although conducing to economy, can hardly effect so great a saving as one making use of gases which otherwise would be wasted. A Schmidt superheater using waste gases is also made.

Mechanical Stokers.—In cases where several boilers are installed, or where there is a single boiler the output from which is fairly constant, it is an advantage to fit each boiler with a mechanical stoker. This apparatus enables the number of men stokers in a large power-house station to be reduced; but what is of much greater importance is the fact that a good mechanical stoker allows a very cheap quality of coal to be used and burnt smokelessly. Small coal known as slack, which, without a mechanical stoker is comparatively useless for raising steam, is now very largely used; it contains from 10,000 to 12,000 B.T.U. per lb., and can be obtained at prices varying from three shillings and sixpence to ten shillings per ton, the price varying with the locality and with the demand, or otherwise, for this very small coal. A mechanical stoker adds a small quantity of coal to the fire at regular intervals, and thus ensures more perfect combustion than is possible when a mass of coal is thrown on at a time.

There are various kinds of mechanical stokers; the difference between each consists chiefly in the manner in which the coal is put on to the fire. In one, the Bennis, the coal is thrown on in the form of spray; this is called a sprinkling stoker. In another, the Vicars, the small coal is pushed forward by means of plungers; this form of stoker is called a coking stoker, as the small coal gets coked at a very early stage in the process of combustion.

Fig. 26.—"Underfeed" stoker.

Another form is the underfeed stoker. In this stoker the centre of the fire grate is higher than the sides; there is an opening running down the whole length of the grate through which coal is fed up from below by means of a worm; the bars fall away on each side of the opening and air is driven in between them. Fig. 26 shows a section through the furnace and bars.

In this form of stoker the coal gets coked early and very perfect combustion is obtained. In most stokers of the above types there is a device for rocking the fire bars, so that the burning fuel is gradually carried forward until it reaches the end of the grate, when it falls over in the form of clinker and ash. In the Bennis stoker a small blast of steam is employed to force air into the furnace. Air for the underfeed stoker is driven in by a fan.

Perhaps the best and simplest form of mechanical stoker is that of the chain grate type. In this stoker the fire bars are connected together so as to form an endless chain, which is moved slowly over drums, thus carrying the fuel forward in a very regular and even manner. An advantage which this stoker has over many others is that the grate may be of unlimited length. Until recently the very smallest form of slack could not be used with chain grate stokers, but the introduction of a new form of fire bar has overcome this difficulty.

An objection which is sometimes urged against a mechanical stoker is that it is somewhat difficult to force a boiler so fitted, should an exceptional quantity of steam be required in an emergency. If, however, the stoker is combined with some system of blast or forced draught, or if the speed of the chain grate can be accelerated or retarded, the objection does not hold good.

Although not necessarily connected with mechanical stokers, two well-known systems of increasing the draught in a furnace may be mentioned. They are the Howden system (as used chiefly in marine work) and the Meldrum blower system. In the Howden system the air is heated by being passed through pipes placed in the uptake from the boiler; the heated air is then supplied under pressure both above and below the firegrate. In the Meldrum system steam is superheated in a pipe placed in the furnace, and is then discharged through a trumpet-shaped blower below the furnace; by this system very small coal can be burnt smokelessly and without a mechanical stoker.

Coal Conveying Plant.—The large number of boilers in a power-generating station, and the use of mechanical stokers, render it almost essential that the coal should be conveyed to the hoppers of the stokers by mechanical means. This is usually effected by means of elevators, conveyers, and occasionally by a crane and grab. The coal bunkers, as a rule, are placed above the boilers, and shoots convey the coal from these bunkers to the hoppers of the stokers.

An elevator consists of a series of pressed steel buckets, about 12 inches wide, which are attached to chains passing over drums; as the buckets travel round the lower drum they come in contact with the coal, either in a barge, bunker, or other receptacle, and fill themselves; when the buckets reach the top of the elevator they tip over and discharge their contents into a shoot leading to a conveyer.

A conveyer may be either of the bucket or chain type; in the latter the conveyer is merely a steel trough, usually from 9 to 18 inches wide, and from 6 to 9 inches deep. At the bottom of this trough a flat chain resembling a ladder moves slowly along; the links of the chain are about 2 or $2\frac{1}{2}$ inches deep, about $\frac{1}{2}$ inch thick, and spaced about 12 inches apart. As the chain moves along it drags the coal with it; the trough passes over shoots which lead to hoppers on the boilers, and in the trough, directly over each shoot, is an opening with a sliding door; a certain proportion of the coal, as it travels along the trough, falls through these openings, the exact amount being regulated by opening or closing the sliding doors. These conveyers will convey coal up an incline of 30° or at an even greater angle, so that in cases where the coal wharf or siding is at some distance from the boiler-house the coal may be conveyed up a gradual incline to the bunkers on the first floor, and an elevator may be dispensed with.

If it is desired to convey the coal at right angles to the main conveyer, the coal is allowed to drop through an opening in the trough and fall on to a second conveyer. These conveyers are driven either by motors, by shafting from the engine-house, or by a separate engine.

The bucket form of conveyer consists of a series of small buckets attached to two double flat-link chains; the chains are provided with wheels which run on suitable rails. When it is desired to discharge coal from this form of conveyer, the bucket is tipped up by means of a cam. Special fillers are provided to ensure that the coal falls into the buckets only as they pass. A conveyer of the bucket type requires less power to drive it than one of the plain chain and trough type.

Oil Filters.—When the steam from an engine is condensed and is used over and over again for feeding the boiler, it is necessary to remove the oil from it. Oil filters are of two different kinds, one designed to extract the oil from the steam before it is condensed, the other to remove the oil from the condensed steam or water. Of the former type the best known examples are the Baker and the Templer-Ranoe. The principles upon which these work are similar; the steam is expanded into a large chamber, when some of the heavier loose particles of oil are thrown down by gravity. The steam then strikes against baffle plates, and its direction is altered, when more oil is thrown down. In the Templar-Ranoe separator the baffle plates are hollow with small lip-like openings, the oil as it trickles down these baffles enters the openings and is not liable to be again licked up by the steam. Such mechanical oil filters will extract about 98 per cent. of the oil from the steam. In the majority of cases the remaining 2 per cent. is harmless, in others trouble arises from it. One of the most effective filters for extracting the oil from steam after condensation consists of chambers containing sawdust, through which the condensed steam is passed. The sawdust, however, requires to be removed and replaced at somewhat frequent intervals. The only way to remove the whole of the oil from steam is to treat it chemically after condensation. Plants for this purpose are, however, somewhat expensive and occupy a good deal of space.

CHAPTER V.

STEAM PIPES AND VALVES.

Steam Pipes.—The pipes conveying steam from the boilers to the engines or turbines are usually made of wrought iron or mild steel, and are lapwelded, solid-drawn, or riveted. Some pipe makers still prefer wrought iron to steel for making lapwelded pipes, as in their opinion a more reliable weld can be obtained with the former. Solid drawn steel tubes can be obtained up to 10 inches diameter, but they are expensive. Lapwelded tubes are made up to 14 inches diameter. Pipes above 14 inches diameter are usually made of riveted mild steel. Cast-iron pipes should not be used for steam pressures above 80 lbs. per square inch.

The flanges at the ends of wrought-iron and steel pipes are made of mild steel, wrought iron, cast iron, or cast steel. They are usually screwed on to the pipe, the end of the latter being expanded and riveted over into a space or recess left in the flange for the purpose. Wrought-iron flanges are sometimes welded on, but unless the welding is done thoroughly, a screwed and riveted flange is to be preferred. The flanges of riveted pipes of large size are, of course, riveted on. The joint between the flanges, which are faced, consists usually of a soft copper ring or of a brass corrugated ring. Any jointing material, such as asbestos, which may blow out, should be avoided.

Arrangement of Pipes.—In laying out a pipe arrangement, care should be taken to avoid hollows or pockets where water can collect. In cases where a pocket is unavoidable, it should be connected by a drain pipe to a steam trap. These will be described later.

Most engineers are fully alive to the danger resulting from water lying in pipes, should it be necessary for these to dip down from the boiler and rise again to the engines, and great care is usually taken to see that such pipes are free from water before starting the engines, and to keep them well drained while the engines are running; but there is another arrangement frequently met with, which is almost as dangerous, and which is probably responsible for a large proportion of the accidents occurring through the presence of water in steam pipes. The arrangement referred to consists of a long horizontal or slightly sloping length of pipe with a vertical rise at one end, beyond which the engines take their steam. The arrangement is shown by Fig. 27.

The sketch shows a long pipe conveying steam to engines A and B. Any water which condenses in this pipe is carried forward by the passage of the steam to the bend C, where it collects and obstructs the pipe to the extent of perhaps one-half of its area or even more; this obstruction may not be serious so long as the engine supplied from flange A, only, is working, but should the attendant suddenly start up the engine connected to flange B, the sudden rush of steam past the bend C will probably pick up the water lodged there, and carry it forward like a bullet, possibly wrecking the engine at B or some other engine further along the line of piping. If such a vertical rise after a

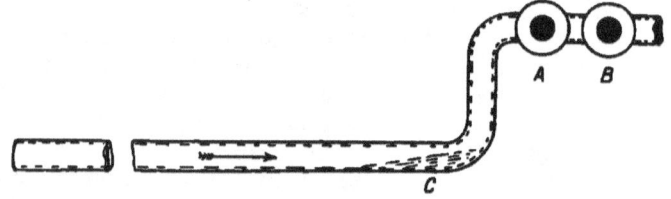

Fig 27.—Dangerous pipe arrangement.

long length of horizontal piping is necessary, the bend should have a large pocket with a drain pipe leading from it to a steam trap.

One occasionally, in past years, received advice to arrange steam pipes with a gradual slope down towards the boiler, so that all the water in the pipes might drain back to it. This advice is radically bad, as a moment's consideration will show, for with the pipes full of steam under pressure, and the steam travelling away from the boiler at a rate of perhaps 6,000 feet per minute, it is practically impossible for any water to travel back against the current of steam. The plan is bad, for another reason, viz. :—When the engine is not working, and the boiler stop valve is closed, any steam condensed in the pipe will flow back towards the boiler, and will lie either on the stop valve, or in the pipes adjacent to it. When the boiler stop valve is next opened, the water collected upon it, unless previously drained off, will have to be got rid of somehow.

The stop valve on the boiler should be arranged as shown by Fig. 28, and not as shown by Fig. 29.

The right way to arrange steam pipes is to erect them with a slight fall all the way from the boilers to the engines, and to provide each engine with a steam separator or dryer. This dryer may be merely a receptacle arranged, so as to change the direction of the flow of steam, and to hold the water as caught. If the direction in which steam is travelling at a high velocity is suddenly altered, any large particles of water entrained in it will be thrown down. The steam dryer should have a gauge glass to show the amount of water lying in it, and be provided with a drain cock connected either to a trap, or to a pipe, leading to a sump or hot well. Any valves on branch pipes leading out of the main steam pipe should be placed close up to the latter, so that water may not collect in the branches above the valves when the latter are closed.

In designing pipe work, another point to be borne in mind is the expansion and contraction of the pipes due to the difference of temperature when full of steam, and when empty and cold. Wrought iron expands ·0000067 times its own length for every degree Fahr. increase of temperature between 32° and 212°, and as much as ·0000089 times its own length for every degree Fahr. when the range of temperature is between 32° and 500°, so that if we have a pipe 50 feet long

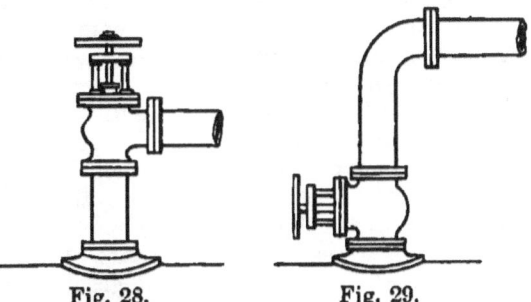

Fig. 28.   Fig. 29.

Positions of stop valve on boiler.

= 600 inches, and its temperature rises from 50° to 480° (which temperature is reached with steam at 185 lbs. pressure and 100° of superheat), we have a difference of 430°. The expansion, therefore, is 600 × 430 × ·0000089 = 2·29 inches, a total expansion of 2·29 inches in 50 feet of pipes.

It is usual to provide for such expansion by having easy wrought-iron or mild steel bends at each end of the pipes, and easy bends where the pipes join and leave the main steam range. Formerly copper bends were used to take up the expansion, but as copper rapidly loses its strength under high temperature, the use of such copper bends has been practically abandoned.

Where it is impossible to provide bends to take up the expansion, an expansion joint, consisting of one pipe running into another of larger diameter, and provided with a gland, is employed; the gland is packed with asbestos, or other packing, in the same way as the gland of a piston-rod is packed.

In fixing steam pipes they should be supported on rollers, or slung in such a way that the contraction and expansion of the pipes may not strain any of the joints.

When electric-generating stations were first installed in this country, it was customary for some engineers to employ duplicate steam mains, in order to guard against any interruption to the supply of current, in the event of the failure of a pipe or pipe joint. This arrangement necessitated a very large number of valves, which were in themselves a source of weakness; and in any duplicate system there is one point where the two pipes converge into a Y-piece, usually on the separator of the engine. Should the joint of this Y-piece blow out, the elaborate duplicate system is rendered useless. It is now more usual to have one large pipe into which the boilers feed, and from which the engines or turbines take their supply. This pipe may be divided into sections by means of full-way valves, if desired.

Size of Pipes.—The size of steam pipes should be such that the flow of steam in them does not exceed a speed of from 5,000 to 6,000 feet per minute (5,000 feet is preferable for pipes of 4 inches and less in diameter); and if the amount of steam to be carried is known, the area of the pipe can easily be calculated on this basis by simple arithmetic. The quickest way to make the calculation is by the formula given below. It should be remembered that the greater the speed of the steam in the pipes the greater is the drop in pressure at the far end of the pipe, or, expressed in another way, a greater amount of pressure is required to force steam through pipes at a high speed than at a low one. (The resistance varies with the square of the velocity.) If the boiler is capable of producing steam at a pressure higher than is required by the engine, then there is not the same objection to employing small pipes; but if the engine, in order to develop its full power, requires steam at approximately the same pressure as that generated in the boiler, then pipes of ample size should be provided.

The formula for ascertaining the area of a pipe to convey a certain number of pounds of steam at a given speed is as follows:—

$$\frac{P \times V \times 144}{S} = A \; ;$$

where P = pounds of steam per minute.
  V = volume in cubic feet of 1 lb. of steam at the pressure to be employed (given in Table IX.)
  S = speed permitted in feet per minute.
  A = area of pipe in square inches.

Example.—What size of pipe is required to convey steam to a turbine using 54,000 lbs per hour, or 900 lbs. per minute, assuming the rate of flow in the pipe is not to exceed 6,000 feet per minute, the steam pressure being 150 lbs.? The calculation is —

$$\frac{900 \times 2\cdot 71 \times 144}{6,000} = 58\cdot 5 \text{ area.}$$

From a table of areas, it will be seen that a pipe 8⅝ inches diameter has an area of 58·42, so that a pipe of this size would suffice; but wrought iron and steel pipes are not made in odd sizes, so that a pipe of the nearest even size—viz., 9 inches—would be selected.

If the area of a pipe is known, and it is desired to ascertain how many cubic feet of steam will flow through it per minute at a given linear speed, the calculation is, of course, as follows:—

$$\frac{A \times S}{144} = \text{cubic feet.}$$

Example.—How many cubic feet of steam will flow through a pipe 6 inches diameter (the area of which is 28·27 square inches), if the speed of the steam is not to exceed 6,000 feet per minute?

$$\frac{28\cdot 27 \times 6,000}{144} = 1,177 \text{ cubic feet.}$$

If the number of cubic feet of steam is known, it can be converted into pounds of steam by the aid of Table IX.

The following table, giving the amount of steam in cubic feet per minute, and in pounds per hour, which will pass through pipes of various sizes when the speed is limited to 6,000 linear feet, may be useful. The approximate loss of pressure for every 100 feet of straight pipe is also given:—

TABLE XI.—Rate of Flow, 6,000 Feet per Minute.

Diameter of pipe.	Cubic feet per minute.	Steam at 150 lbs. (gauge) pressure.		Steam at 200 lbs. (gauge) pressure.	
		Lbs. per hour.	Approximate drop of pressure per 100 feet.	Lbs. per hour.	Approximate drop of pressure per 100 feet.
Ins.			Lbs. per sq. in.		Lbs. per sq. in.
3	294	6,514	4·0	8,320	5·25
4	523	11,579	3·0	14,802	4·0
6	1,177	26,059	2·0	33,311	2·6
9	2,650	58,671	1·4	75,000	1·75
12	4,708	104,236	1·0	133,245	1·3
18	10,600	234,686	·6	300,000	·87

The number of cubic feet of steam per minute that will flow through a pipe of a given size and at a given speed in linear feet per minute is, of course, the same, whatever the working pressure of the steam may be; but as the density of steam is greater at high than at low pressures, a larger number of lbs. of high-pressure steam will come through than would be the case with low-pressure steam; the pressure or head required to drive the denser steam through is, however, greater.

Thus, if we take the case of a 6-inch pipe, in which the steam flows at a rate of 6,000 feet per minute, 26,059 lbs. of steam at 150 lbs. pressure will be delivered with a fall of pressure of about 2 lbs. in each 100 feet of straight length of pipe; but with steam at 200 lbs. pressure, 33,311 lbs. will come through, but the fall of pressure will be about 2·6 lbs. per 100 feet, the number of cubic feet coming through being 1,777 in each case.

The generally-accepted formula dealing with the flow of steam in pipes is given below,* but this formula is interesting as showing what the steam should do in smooth pipes, rather than what it actually does in practice.

In actual working practice there are many factors which may entirely vitiate the results of calculations based on elaborate formulæ which cannot well take these factors into account. For instance, in a long range of straight steam pipe there is, at a distance of every 12 feet or so, a tiny groove formed by the space required for the packing placed between the flanges. This packing ring cannot be made of the same internal diameter as the pipe, on account of the danger of a part of the packing projecting into the bore of the pipe and thus causing an obstruction. These tiny grooves cause eddies in the steam and tend to impede its flow. Again, in erecting the pipes one cannot be sure that the centre of each length of pipe will coincide absolutely with that of its neighbour. There is a certain amount of clearance in the flange bolt holes, and it is hardly probable that all the pipes in a long range are absolutely true one with another. The wetness or dryness, too, of the steam affects its rate of flow.

The obstruction to the flow of steam caused by a globe valve

$$* W = 87 \sqrt{\frac{D(p_1 - p_2) d^5}{L\left(1 + \frac{3·6}{d}\right)}}$$

W = weight in lbs. per minute.
D = weight per cubic foot of steam at the working pressure.
p_1 and p_2 = the initial and final pressures.
L = length of pipe in feet.
d = diameter of pipe in inches.

is sometimes stated to be equivalent to so many feet of straight pipe, varying from 50 feet in a 3-inch pipe to 80 feet in a 12-inch pipe, but if the globe valve is not drained, and water accumulates in it as shown by Fig. 30, no formula will indicate what obstruction is caused.

Even if properly drained, the roughness and shape of the valve body will affect the rate of flow considerably.

A method sometimes adopted by a draughtsman to settle the size of steam pipes, is to ascertain the diameter of the inlets upon all the engines to be supplied with steam, and which will be working at the same time, and then to arrange for his steam pipe to have a corresponding area. For instance, should there

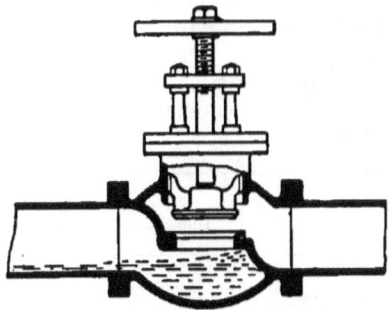

Fig. 30.—Stop valve passage obstructed by water.

be two engines each with an inlet of 6 inches diameter, and one with an inlet of 5 inches diameter, the total area of these inlets amounts to 76·1 square inches; the draughtsman turns to a table of areas and finds that a pipe 10 inches diameter has an area of 78·5 square inches; he then settles on this size of pipe. If there are a large number of engines this method gives a size of pipe larger than is really needed, as the engine maker, in settling the size of his inlet, usually allows a little margin.

The principal objection to the use of pipes of unduly large size, is the heavy first cost, both of pipes and valves. There

---

A good empirical formula for giving the flow of steam in pipes is as follows:—

$$S = 50\sqrt{\frac{V \times F \times 144 \times D}{L}};$$

where S = speed (in linear feet) of the steam per second.
V = volume in cubic feet of 1 lb. of steam at the working pressure.
F = fall in pressure.
D = diameter of the pipe in feet.
L = length of the pipe in feet.

is also a greater loss from radiation with a large pipe than with a small one. It may be mentioned that the loss from this source is about 13 or 14 B.T.U. per square foot of an uncovered pipe. This loss may be reduced to one varying from 1½ to 4 B.T.U. per square foot if the pipe is covered with one of the various non-conducting compositions sold for the purpose.

On the other hand, a large steam pipe provides a certain reservoir of steam; this is an advantage when the steam is required for a slow-speed reciprocating engine, which demands a large supply of steam while the admission port is open, and requires no steam while the port is closed. This matter will be referred to again under valves.

The effect of superheating steam is to increase its volume, not its pressure, but superheated steam travels along a pipe with less friction than wet steam and prevents any obstruction due to water, such as is shown by Fig. 30.

The increase of volume caused by superheating steam is approximately as follows:—

	100° Superheat. Increase of volume.	150° Superheat. Increase of volume.
100 lbs. pressure, . . .	12·5 per cent.	19 per cent.
150 ,, ,, . . .	12·0 ,,	18 ,,
200 ,, ,, . . .	11·8 ,,	17 ,,

If, therefore, pipes are designed so that a certain weight of saturated steam, at 200 lbs. pressure, should flow at 6,000 feet per minute, then the same weight of steam superheated by 100° will require to flow at about 6,708 feet per minute. There would be no real objection to this increased rate of flow, as superheated steam approaches more nearly to a perfect gas than saturated steam, and its friction is less.

Strength of Pipes.—The tensile stress in the material of a pipe or cylinder under internal pressure is found by the following formula:—

$$S = \frac{D \times P}{2 \times T};$$

where S = Tensile stress per square inch.
D = Diameter of pipe in inches.
P = Pressure in pounds per square inch.
T = Thickness of pipe in inches, or parts of an inch.

In the case of lapwelded wrought-iron and steel pipes, it is not customary to allow a stress equal to the safe working stress

STEAM PIPES AND VALVES.

of the metal, as the strength of the weld and the stiffness of the pipe generally have to be taken into account.

The usual thickness of wrought-iron and steel pipes suitable for working up to pressures of 200 lbs. per square inch is as follows:—

TABLE XII.

Diameter of Pipe.	Lapwelded.		Solid Drawn Steel.
	Wrought Iron.	Steel.	
Inches.	Inches.	Inches.	
3	$\frac{1}{4}$	$\frac{1}{4}$	About $\frac{1}{32}$ less in thickness than lapwelded steel tubes.
4	$\frac{1}{4}$	$\frac{1}{4}$	
5	$\frac{1}{4}$	$\frac{1}{4}$	
6	$\frac{5}{16}$	$\frac{1}{4}$	
7	$\frac{6}{16}$	$\frac{1}{4}$	
8	$\frac{6}{16}$	$\frac{5}{16}$	
9	$\frac{6}{16}$	$\frac{5}{16}$	
10	$\frac{3}{8}$	$\frac{5}{16}$	
12	$\frac{3}{8}$	$\frac{3}{8}$	
14	$\frac{7}{16}$	$\frac{3}{8}$	

If the reader will apply the formula given above to the pipes in the table, he will find that with a working pressure of 200 lbs. the stress in the metal is only about $\frac{1}{2}$ ton per square inch in the case of the 3-inch pipe; below 1 ton in the case of the 4-inch, 5-inch, and 6-inch pipes; and less than $1\frac{1}{2}$ tons in the case of the 12-inch pipe.

As in practice the metal is stressed to such a small extent there is but little advantage in employing mild steel for lapwelded pipes if a better weld can be obtained, which is doubtful, with wrought iron.

Lapwelded pipes of less than 3 inches diameter, whether for steam or gas, are made of a uniform outside diameter—this diameter will be found in the table of gas threads (Table v., Chap. II.)—and the inside diameter is made greater or smaller according to the pressure the pipe is required to withstand. Thus a 2-inch steam pipe is of a slightly smaller internal diameter than a 2-inch gas pipe, the outside diameter being the same. This rule does not apply to hydraulic tubes, which are of considerable thickness.

Size of Flanges.—Until quite recently, every engineer has used his own standard dimensions for pipe flanges. This has caused a great deal of inconvenience, especially to valve makers.

Recently, however, the Engineeering Standards Committee, a body composed of members of the Civil and Mechanical Engineers' Institutions, of the Institution of Naval Architects, and of the Iron and Steel Institute, drew up a table settling the diameter and thickness of flanges, the number of bolts to be used, and the pitch circle; both for steam pipes suitable for pressures up to 225 lbs. per square inch, and for exhaust pipes, or pipes suitable for working up to pressures not exceeding 55 lbs. per square inch.

These tables are appended and may prove useful; many engineering firms have already adopted them, and others intend to do so as soon as existing stock has been disposed of.

TABLE XIII.—Dimensions of British Standard Pipe Flanges.

For steam pressure up to 225 lbs. per square inch.

Internal diameter of pipe.	Diameter of Flange.	Diameter of Bolt Circle.	Number of Bolts.	Diameter of Bolts.	Thickness of Flange.	
					Cast iron or Wrought Iron.	Steel or Bronze.
Inches.	Inches	Inches.				
1	4¾	3 7/16	4	5/8	5/8	1/2
1¼	5¼	3⅞	4	5/8	5/8	1/2
1½	5¼	4⅛	4	5/8	3/4	9/16
2	6½	5	4	5/8	7/8	11/16
2½	7¼	5¾	8	5/8	7/8	11/16
3	8	6¼	8	5/8	1	3/4
3½	8½	7	8	5/8	1	3/4
4	9	7½	8	5/8	1⅛	7/8
5	11	9¼	8	3/4	1¼	1
6	12	10¼	12	3/4	1¼	1
7	13¼	11½	12	3/4	1⅜	1⅛
8	14½	12¾	12	3/4	1⅜	1⅛
9	16	14	12	7/8	1½	1¼
10	17	15	12	7/8	1½	1¼
12	19¼	17¼	16	7/8	1⅝	1⅜
14	21¾	19½	16	1	1¾	1½
16	24	21¾	20	1	1⅞	1⅝
18	26½	24	20	1⅛	2	1¾
20	29	26½	24	1⅛	2⅛	1⅞
24	33½	30¾	24	1¼	2⅜	2⅛

TABLE XIV.—Dimensions of British Standard Pipe Flanges.

For steam pressures up to 55 lbs. per square inch.

Internal diameter of pipe.	Diameter of Flange.	Diameter of Bolt Circle.	Number of Bolts.	Diameter of Bolts.	Thickness of Flanges.	
					Cast Iron.	Steel or Bronze.
Inches.	Inches.	Inches.				
1	$4\frac{1}{2}$	$3\frac{1}{4}$	4	$\frac{1}{2}$	$\frac{1}{2}$	$\frac{3}{8}$
$1\frac{1}{4}$	$4\frac{3}{4}$	3	4	$\frac{1}{2}$	$\frac{5}{8}$	$\frac{1}{2}$
$1\frac{1}{2}$	$5\frac{1}{4}$	3	4	$\frac{1}{2}$	$\frac{5}{8}$	$\frac{1}{2}$
2	6	$4\frac{1}{2}$	4	$\frac{5}{8}$	$\frac{3}{4}$	$\frac{9}{16}$
$2\frac{1}{2}$	$6\frac{1}{2}$	5	4	$\frac{5}{8}$	$\frac{3}{4}$	$\frac{9}{16}$
3	$7\frac{1}{4}$	$5\frac{3}{4}$	4	$\frac{5}{8}$	$\frac{3}{4}$	$\frac{9}{16}$
$3\frac{1}{2}$	8	$6\frac{1}{2}$	4	$\frac{5}{8}$	$\frac{3}{4}$	$\frac{9}{16}$
4	$8\frac{1}{2}$	7	4	$\frac{5}{8}$	$\frac{7}{8}$	$\frac{11}{16}$
5	10	$8\frac{1}{4}$	8	$\frac{5}{8}$	$\frac{7}{8}$	$\frac{11}{16}$
6	11	$9\frac{1}{4}$	8	$\frac{5}{8}$	$\frac{7}{8}$	$\frac{11}{16}$
7	12	$10\frac{1}{4}$	8	$\frac{5}{8}$	1	$\frac{3}{4}$
8	$13\frac{1}{4}$	$11\frac{1}{2}$	8	$\frac{5}{8}$	1	$\frac{3}{4}$
9	$14\frac{1}{2}$	$12\frac{3}{4}$	8	$\frac{5}{8}$	1	$\frac{3}{4}$
10	16	14	8	$\frac{3}{4}$	1	$\frac{3}{4}$
12	18	16	12	$\frac{3}{4}$	$1\frac{1}{8}$	$\frac{7}{8}$
14	$20\frac{3}{4}$	$18\frac{1}{2}$	12	$\frac{7}{8}$	$1\frac{1}{4}$	1
16	$22\frac{3}{4}$	$20\frac{1}{2}$	12	$\frac{7}{8}$	$1\frac{1}{4}$	1
18	$25\frac{1}{4}$	23	12	$\frac{7}{8}$	$1\frac{3}{8}$	$1\frac{1}{8}$
20	$27\frac{3}{4}$	$25\frac{1}{4}$	16	$\frac{7}{8}$	$1\frac{1}{2}$	$1\frac{1}{4}$
24	$32\frac{1}{2}$	$29\frac{3}{4}$	16	1	$1\frac{5}{8}$	$1\frac{3}{8}$

Water Hammer.—When steam at a high pressure is admitted to a pipe in which a certain amount of cold water is lying, a succession of sharp reports is heard, just as though blows were being struck on the inside of the pipe by a hammer; sometimes the blows are sufficiently strong to fracture the pipe or a neighbouring valve.

The precise action of water hammer has recently been investigated by Mr. Strohmeyer, who used glass tubes for his investigations. Mr. Strohmeyer found that when steam was admitted to a pipe containing water, waves were set up in the latter which imprisoned and isolated portions of the steam. The isolated portion of the steam being in contact with cold water, and no further supply of heat being able to come to its rescue, condensed, the water then rushed into the space formerly occupied by the steam and caused a sharp blow.

In addition to this action of water hammer proper, the presence of water in a steam pipe, as explained earlier in this chapter and illustrated by Fig. 27, may lead to disaster. In such a case the water is picked up by the rush of steam and acts as a projectile.

Sometimes an accident is caused through the attendant not realising what goes on in a steam pipe after the engine and boiler valves have been closed. What happens is this:—After the valves have been closed for some time, and heat has been lost by radiation, the steam condenses and forms water; if the valves are tight no air can get into the pipes and a vacuum is formed. Now, if the drain cocks on the pipe are opened before a small quantity of steam is admitted, air will pass into the pipes, but water will not come out against atmospheric pressure. The attendant, seeing no water coming from the pipe, may conclude that it is free from water, and may turn on the steam too suddenly, with disastrous results. At an enquiry recently held by the Board of Trade upon an accident due to water hammer, the attendant said that when he opened the cocks he saw no water coming from the pipes, but heard a hissing noise; even this did not convey to him the fact that there was a vacuum in the pipes, and that the hissing sound he heard was due to air rushing in.

Exhaust Pipes are usually made of cast iron, as the pressure they have to withstand is low; when the engine is exhausting to the atmosphere the pressure in the pipes is not much above that of the atmosphere. When the engine is condensing the pipe is subject to a crushing stress not exceeding 15 lbs. per square inch.

In calculating the strength of a wrought-iron pipe, we said that the weld and general stiffness of the pipe had to be taken into consideration. In a cast-iron pipe there is of course no weld, but in casting the pipe the core may have shifted, and one side of the pipe may be much thinner than the other. Even if the pipe were of the same thickness throughout, a pipe calculated on the basis of the safe stress for cast iron would be much too thin, especially in cases where the pressure is low. In practice the thickness of exhaust pipes is approximately as follows:—

Diameter.	Thickness.	Diameter.	Thickness.
Inches.	Inch.	Inches.	Inch.
3	$\frac{3}{8}$	9	$\frac{1}{2}$
4	$\frac{3}{8}$	10	$\frac{9}{16}$
5	$\frac{5}{8}$	12	$\frac{5}{8}$
6	$\frac{7}{8}$	18	$\frac{3}{4}$
8	$\frac{1}{2}$	20	$\frac{7}{8}$

STEAM PIPES AND VALVES. 83

Steam Traps are devices for allowing water automatically to leave a pipe containing steam under pressure without permitting steam to pass. Steam traps are of two types—viz., the bucket and expansion types. The former depends for its action upon

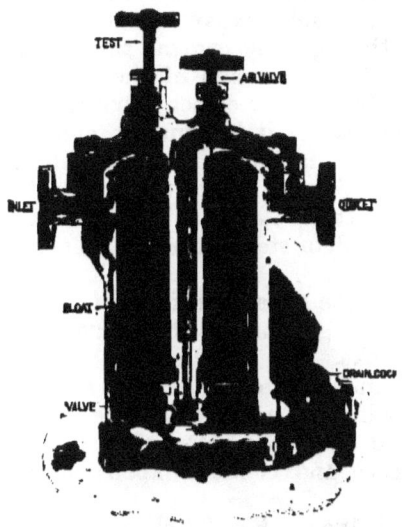

Fig. 31.—Bucket trap.

a floating bucket; the latter depends upon the difference of temperature between the water which has condensed and that of the live steam. Fig. 31 shows a bucket steam trap as made by Messrs. Holden & Brooke, of Manchester. At starting, the trap is partly filled with water, which causes the floating bucket

Fig. 32.—Expansion trap.

to rise and close the outlet valve. When water comes down from the steam pipe it gradually fills the shell, reaches the top of the bucket, and flows into it; when the bucket is full its buoyancy has, of course, disappeared, and the weight of the bucket is sufficient to open the outlet valve. The pressure of steam on the surface of the water then drives the latter out of the bucket, up the central outlet pipe; when it has been expelled the weight of water surrounding the bucket causes the latter to rise and to again close the outlet valve.

Fig. 32 shows a Brooke expansion trap. In this trap steam can blow freely through the central tube until the steam has warmed it, when the expansion of the tube causes the outlet valve to close. When water comes down from the steam pipe it cools the tube and causes it to contract, consequently the valve opens and the water is blown out. A difference of 5° F. in temperature is sufficient to make the trap act; in fact, it is so sensitive that if a little water is sprinkled off the hand on to the central tube when full of steam, the valve opens, spits out steam, and then closes.

The objection to the expansion form of trap is that the valve and seat after a time get cut by the outgoing steam and water, and the valve leaks. In the Brooke bucket trap the outlet valve is always under water, and is given a rotary motion, so that it grinds itself in at every discharge. The bucket form of trap is considered the more reliable of the two kinds, but it takes up more space, and is more expensive than the expansion form of trap.

Steam Stop Valves.—Fig. 33 shows a screw-down, or globe right-angled boiler stop valve, having a renewable seating; that is to say, the valve seat is not made solid with the valve body, but a separate metal seat is forced in to the body, and is held in by set screws. The reason for making the seat renewable is that a valve seat, especially if the valve is used for regulating the flow of steam, gets scored or cut by the action of the steam, and the valve is no longer tight when closed.

In the stop valve illustrated, the valve and seat can be renewed when scored without necessitating the renewal of the body. The valves and seats are usually made of some alloy which does not corrode; gunmetal was formerly used, but this alloy is not suitable for superheated steam. One firm of valve makers (Messrs. Templer & Ranoe) use a special nickel alloy for their valves and seats, while another firm (Messrs. Hopkinson) use an alloy which they call platnam; this must not be confused with platinum. Platnam is doubtless a fancy name. The bodies of valves for high-pressure steam should be made of cast steel and not of cast iron.

Valves of the screw-down pattern should be arranged so that the steam assists the valve to open, and constructed so that the screw forces the valve on to its seat; if a valve is constructed so that the spindle draws the valve on to its seat, it is difficult to keep the valve tight.

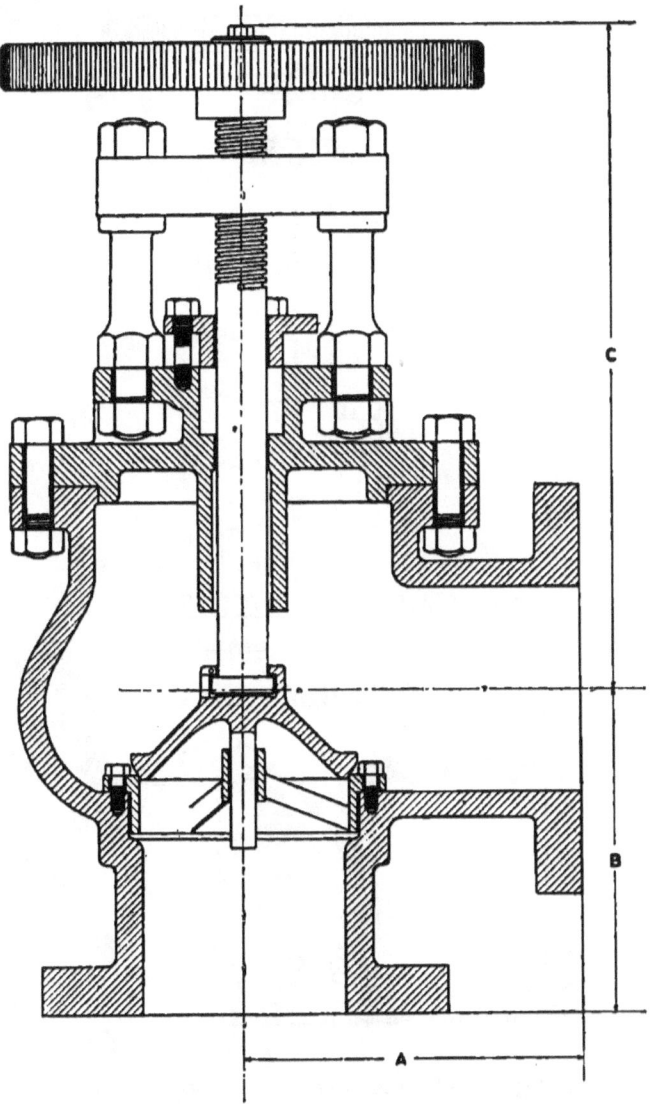

Fig. 33.—Screw-down stop valve.

Fig. 34 shows a very good form of valve of the gate, or straight-through type. The valve shown is known as the Stirling (made by Templer & Ranoe). It will be seen from the illustration that when the valve is closed the wedges force the two

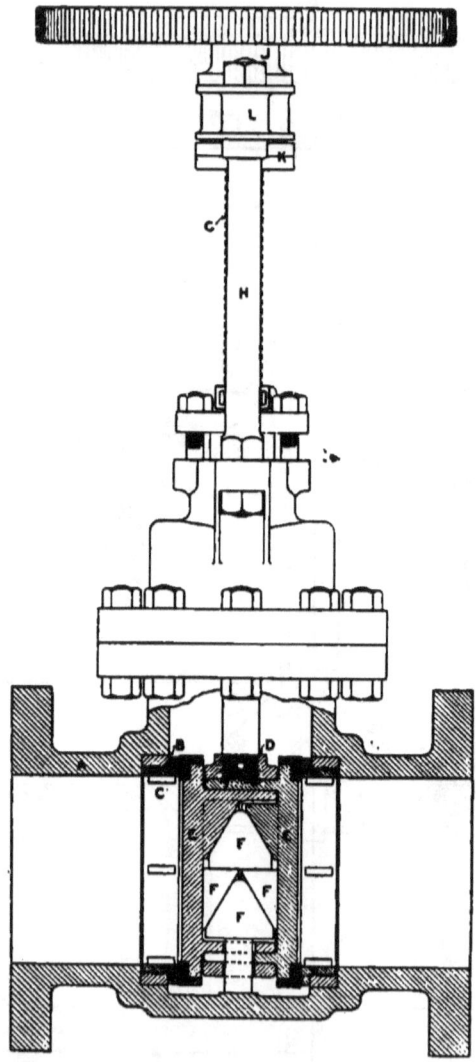

Fig. 34.—Straight-through gate stop valve.

valve faces outwards against the seatings. The two wedges ensure the upper and lower portions of the valve being forced equally against the seats. The Stirling is a double-faced valve—

i.e., both valve faces make a steam-tight joint. In valves of somewhat similar type, suitable for water and for exhaust steam, and known as sluice valves, one face only of the valve frequently makes the joint, but double-faced sluice valves are also made.

It is usual to provide large valves of the gate type with a small by-pass valve, so that the pressure on both sides of the large valve may be equalised before opening it. This by-pass valve is useful for admitting a small quantity of steam to the pipes to warm them up, before opening the main valves.

Valves of the screw-down globe pattern are generally used upon boilers or engines where it may be necessary to use the valve for regulating the flow of steam. Valves of the gate type are generally used in the range of pipe to shut one portion off. A valve of the gate or straight-through type causes less obstruction to the passage of the steam than one of the globe type.

Various forms of valve have been designed to get over the cutting action of the steam; they usually take the form of two valves in one body, one valve opening before steam can pass the second; the first valve therefore does not get cut by the action of the steam. It should be explained that the cutting action does not take place when the valve is fully open, but chiefly when opening, or when the valve is kept partially closed for regulating purposes. The "T. R." valve (made by Messrs. Templer & Ranoe) was probably the first valve of this kind; it is rather expensive, but remains tight for a very long time Messrs. Hopkinson also make a valve of a somewhat similar kind which is called the "centre pressure" valve.

Hopkinson-Ferranti Valve.—A steam valve, which has recently been introduced and largely advertised, is the Hopkinson-Ferranti valve. The bore of this valve is gradually reduced by a nozzle-shaped body, until the bore is only half the diameter of the pipe to which the valve is attached. The valve itself is, of course, at the smallest portion of the bore; beyond the valve proper the body gradually opens out in a suitably proportioned manner until the full area of the pipe is again reached. The principle upon which the valve is constructed, as given by the makers, is as follows:—"Converting the pressure of the fluid into velocity and reconverting the velocity into pressure, thereby passing an amount of steam equal to the full carrying capacity of the pipe." The makers claim that the valve is lighter and cheaper than one having the full opening of the pipe, that there is less risk of leakage, and that the valve causes practically no drop in the pressure of the steam passing through it. The author learns, from independent sources, that when a single valve of the kind is employed the drop of pressure is almost negligible, provided the valve is fully opened; but if the attendant should fail to open the valve to the very fullest extent, so as to cause the slightest obstruction to the steam where it flows through the throat of the valve at a high velocity, then the drop of pressure is serious.

It can hardly be seriously contended that there is no loss in the conversion of pressure into velocity and reconverting velocity into pressure. It is difficult to imagine steam flowing through a long corrugated pipe, however accurately the corrugations may be proportioned, as freely as through a plain pipe. Joule's law, given in the chapter on steam turbines, reads—"When a gas expands without doing external work, and without taking in or *giving out heat*, its temperature does not change." We know that when steam is passed through a constricted opening it becomes slightly superheated and expands, and it is difficult to see how the makers can prevent the steam from giving up some of this superheat to the valve, and so losing it by conduction and radiation.

Apart from theoretical considerations there is one point which should not be overlooked by those using this form of valve. Every length of horizontal pipe placed between two such valves should be properly drained, otherwise a pocket is formed in which water can lie.

Isolating Valves.—Another form of valve used on boilers in power-generating stations is the isolating valve. This valve will allow steam to leave a boiler, but will not allow any to re-enter. Thus if several boilers deliver steam into one common steam main, and one boiler should develop a serious leak, or burst a tube, that boiler only will be put out of action, if fitted with an isolating valve, as this valve will prevent steam entering from the other boilers. Isolating valves sometimes give trouble by hammering and breaking. This trouble is often experienced when the boilers supply steam to slow-speed reciprocating engines, and when the pipes are of small size. The explanation is doubtless as follows:—When the steam port of the big slow-speed engine opens steam travels along the pipe at a very high velocity, the cut off then suddenly takes place, and the flow of steam at one end of the pipe is checked; the steam, which was in motion in the pipes, banks itself up, so to speak, and the pressure rises higher than that in the boiler. Steam then flows back to the boiler, and closes the isolating valve with a sharp blow. Such a blow repeated sixty or seventy times a minute naturally causes the valve to collapse in a short time.

CHAPTER VI.

THE STEAM ENGINE.

(Part I.)

It is beyond the scope of this book to describe in detail a great variety of steam engines; all that is attempted is to make clear the principles upon which such engines work, and to show how the ordinary calculations connected with them are made.

Fig. 35 shows a vertical double-acting engine with cylinder and valve chest in section. The valve shown is of the plain **D** type, as it is necessary to understand the action of this form of valve before considering valves of the piston or other types. The action of the engine is this:—Steam from the boiler enters through an inlet at the back of the valve chest A; the slide-valve S, in the illustration, is just beginning to uncover the steam port leading from the valve chest to the upper side of the piston B; the valve will continue to move downwards until the port is fully open, when it will begin to move in an upward direction. The pressure of steam on the piston B causes it to descend and turn the crank C and crank shaft D towards the spectator by means of the piston-rod E and connecting-rod F.

On the crank shaft is fitted an eccentric H, which gives motion to the valve by means of the eccentric-rod J, and valve-rod K. The eccentric is set in such a position, and the valve is so proportioned, that when the piston has reached about three-quarters of its downward stroke the slide valve will have travelled upwards sufficiently far to prevent the admission of any more steam. This is called the point of cut-off; after the cut-off has taken place, the steam which is above the piston expands and forces the piston down to the end of its stroke. A flywheel, L, is mounted on the crank shaft, the stored energy of which carries the crank over the dead centres (*i.e.*, the position in which the piston is at the extreme end of the stroke), by which time the slide valve will be in such a position that it is just opening the port leading to the underside of the piston; the slide valve will continue to open the port and the piston will rise; at three quarters of its upward stroke steam will be cut off, and the expansion of the steam beneath it will cause the piston to complete its stroke.

It will be seen from the illustration that when one port is uncovered to admit steam, the slide valve places the other port

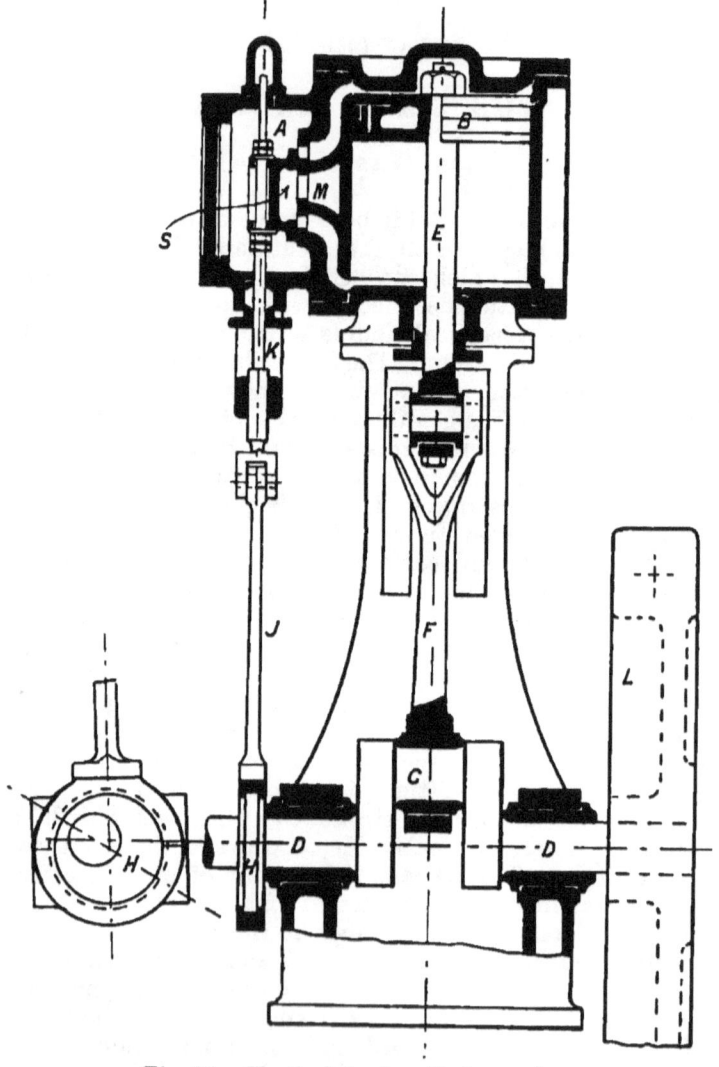

Fig. 35.—Vertical single-cylinder engine.

in communication with the central port or passage M, which, in a single-cylinder engine, leads to the atmosphere or condenser, and in a compound engine to the second cylinder.

The port through which the steam exhausts is closed by the valve slightly before the piston reaches the end of its stroke, so that the imprisoned steam acts as a cushion, and helps to bring the piston to a standstill, and to restart it on its reverse stroke. In a condensing engine this cushioning effect is lost to the cylinder placed in connection with the condenser; this accounts for the fact that certain high-speed engines are much more noisy when condensing than when exhausting to the atmosphere.

It may be noticed that the faces of the slide valve are wider than the steam ports; the difference is called the "lap"; the amount by which the valve faces project beyond the ports in an outward direction when the valve is central, is called "outside lap." The amount of outside lap and the position of the eccentric regulate the point of cut-off. If, when the slide valve is in a central position, the valve faces overlap the ports on the inside, it is called "inside lap."

It will be noticed that, although the piston has not commenced its downward stroke, the valve has slightly uncovered the steam port; this is called "the lead of the valve." Thus, if the port is opened by one-eighth of an inch when the piston is still at the highest part of its stroke, the valve is said to have one-eighth inch lead.

By cutting off the admission of steam fairly early, and allowing it to work expansively, considerable economy is effected, but with a single D slide valve, if sufficient lap is given to make the cut-off take place before about two-thirds of the stroke, the exhaust is closed too early, and too much compression results.

With the D-form of slide valve the pressure of steam on the back of the valve is more than sufficient to hold it up against the valve chest face. Thus a small valve, say 5 inches by 4 inches, has an area of 20 square inches, and if the steam pressure in the valve chest is 120 lbs., a pressure of 2,400 lbs., or over a ton, will be exerted on the back of the valve. This great pressure renders the plain slide valve unsuitable for cases where the steam pressure is high, or where the engine runs at a high speed.

In an engine designed for even moderately high speeds or high steam pressures a valve of the piston, or other type to be described later, would be used. A valve of the piston type is shown by Fig. 36. With this form of valve a liner is employed, in which the ports can be cut more accurately than is possible in the valve-chest casting; bars are left, so that, if rings are used in the piston valve, they will not enter the ports.

These bars obstruct, to a certain extent, the area of the port, and to get the required area a somewhat large piston valve is necessary. The piston valve is usually made about half the diameter of the cylinder, and such a large valve is inadmissible in the case of the low-pressure cylinder of a big compound or triple-expansion engine.

When a plain slide valve is used for the low-pressure cylinder

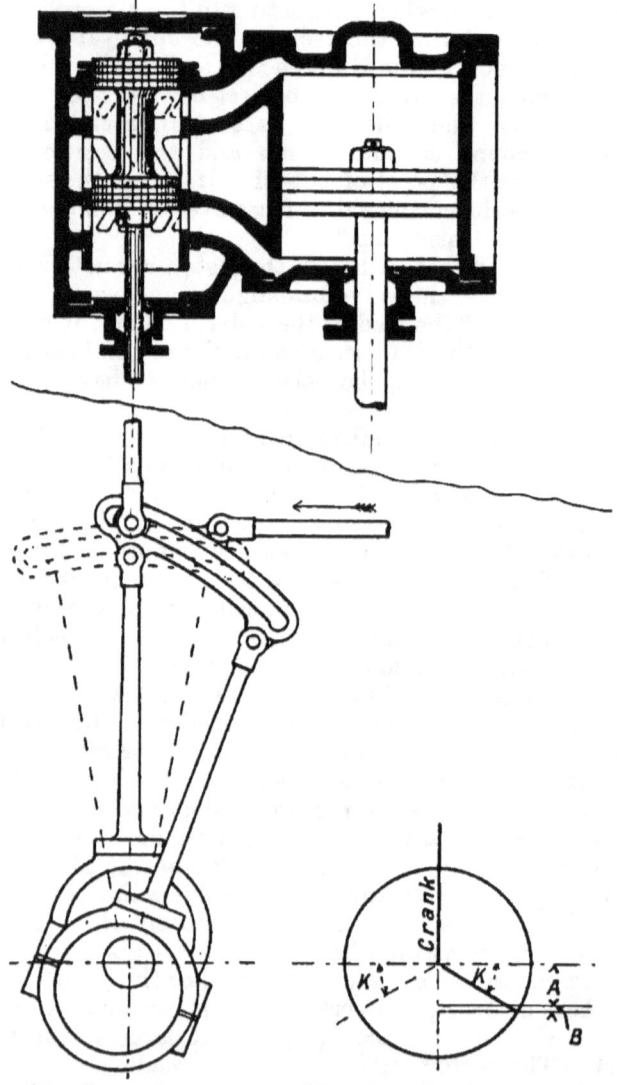

Fig. 36.—Cylinder with piston valve.

Fig. 37.—Reversing gear.   Fig. 37a.—Setting eccentrics.

THE STEAM ENGINE. 93

of a large engine, it is usually balanced—*i.e.*, the valve and valve chest are constructed so that the steam pressure acts upon a small area only of the back of the valve. Fig. 38 shows the low-pressure cylinder of a large vertical engine, for marine or land use, fitted with a double-ported balanced slide valve. In the example shown, steam is prevented from pressing upon a large

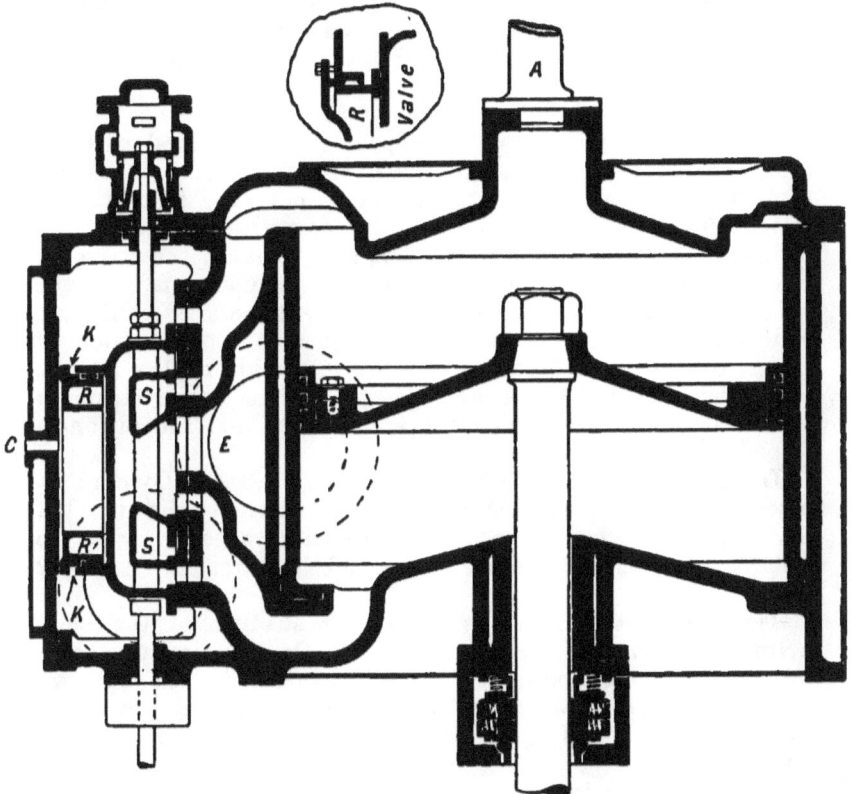

Fig. 38.—Low-pressure cylinder, with balanced slide valve, assistant cylinder, and metallic packing.

portion of the valve by means of the ring R. This ring fits into a circular ring at the back of the valve, and is kept steam tight by means of piston rings. The face of the ring next the valve chest cover and the inside of the latter are planed, the ring is kept tightly up against the valve chest cover by the steam pressure acting at K. Should any steam leak past the ring R, it is taken away to the condenser through the opening C.

There are many other methods of preventing the steam from

pressing upon the back of the valve. In many cases the back of the valve is planed, and the ring R rubs on it. Instead of using piston rings, the ring R is frequently packed by means of a gland, as shown by the inset at the top of the illustration. In the earlier methods of balancing a gland was not employed, the ring merely fitted into a recess in the valve chest cover, the recess being filled with asbestos packing. This was unsatisfactory, for if the fitter put too much packing into the recess, the ring was forced down too hard on to the valve; if too little packing was used, steam leaked by.

In modern marine practice the valve, in addition to being balanced in the manner described, is usually provided with an assistant cylinder placed at the top of the valve chest, as shown. The object of this assistant cylinder is not only to carry the weight of the valve, thus relieving the eccentric straps, reversing link, pins, &c., from the weight, but also to assist the valve to rise, and to force it gently down at the right moments. In the "Joy" assistant cylinder the admission of steam is effected by a reduction in the size of the valve-rod, so that at one portion of the stroke steam is admitted to the underside of the piston; when the piston has reached the top of its stroke steam is admitted to its upper side by means of slots, and the piston is forced gently down.

In another form of assistant cylinder—viz., the "Lovekin"— a separate valve is used to distribute the steam to the cylinder, and steam is taken from the intermediate receiver so as to obtain it at higher pressure, thus enabling a smaller cylinder to be used.

The assistant cylinder is a step in advance of the plain cylinder, which was formerly fitted to the low-pressure valve chests of marine engines. This plain cylinder was open at one end to the valve chest, so that steam pressed constantly on the underside of the piston which carried the main slide valve.

The spaces S S in the slide valve itself are open to the steam in the valve chest, so that the valve is double ported. This construction reduces the amount of valve travel for a given size of port opening. In the illustration the valve is fully open, admitting steam to the upper side of the piston, and placing the lower side in communication with the condenser through the exhaust passage E. A facing piece of hard cast iron, in which the ports are accurately cut, is placed between the valve and the main casting.

The illustration shows the piston- and valve-rods packed by means of metallic packing, which will be described later. The cylinder cover is fitted with a spring relief valve shown at A. The object of this relief valve is to allow any water to escape should such have accumulated on the piston. The piston is provided with a separate ring in which the Ramsbottom piston rings are placed. This admits of the rings being renewed without the necessity of removing the piston from its rod.

Reversing.—When the eccentric is fixed to the shaft in one position, it will allow the engine to rotate in one direction only. If the engine is required to rotate in the opposite direction as well, it is necessary to have two eccentrics, one keyed to the shaft in a position to make the engine run in one direction, the second in a position to make it run in the opposite direction. The rods leading from the two eccentrics are connected to opposite ends of a link. The valve spindle or rod is connected to a block placed in the slotted link, the link itself can be shifted so that the valve-rod may be placed in connection with either of the two eccentrics.

Fig. 37 shows the ordinary Stephenson link motion. One eccentric is in its uppermost position, and owing to the position of the link it is this eccentric which is actuating the valve. It will be seen that the lower port of the steam chest (Fig. 36) is uncovered to the steam, while the upper port is in communication with the exhaust. If now the link is pushed over to the left in the direction of the arrow, the valve will be drawn down and the upper port will be uncovered to the steam, while the lower port will be placed in communication with the exhaust. The shaft in Fig. 37 is at right angles to the position it would occupy if it were working in connection with the cylinder shown by Fig. 36; the two views have been so placed to show clearly the action of the two eccentrics and link, and their relation to the valve.

In actual practice the eccentrics are not set at the highest and lowest positions as shown. They are set at an angle of 90°, plus the angular advance, in advance of, or behind the crank. The actual setting of the eccentrics is shown by Fig. 37a, where the diameter of the circle represents the travel of the valve. A = lap, B = lead. The angle K is called the "angular advance." A drawing similar to that shown by Fig. 37a is sent from the drawing office to the shops with the dimension A + B given, and the diameter of the circle. This enables the fitters to set the eccentrics in the correct positions.

An engine such as that shown by Fig. 35 would not work very economically as regards steam consumption, and would only be used in cases where simplicity and small first cost happened to be of more importance than economy of steam and fuel. Such an engine might be used for driving a fan at a colliery where fuel is cheap, or it might be used in cases where the work is intermittent, and where the somewhat excessive consumption of fuel during the short periods when the engine is at work is of relative unimportance.

A single-cylinder engine working with a steam pressure of 75 lbs. and exhausting to the atmosphere, would use between 30 and 35 lbs. of steam for every indicated horse-power per hour; so that if the engine indicated 50 horse-power, the boiler supplying the steam would have to evaporate between 1,500 and 1,750 lbs. of water per hour.

By *indicated horse-power*, or I.H.P., is meant the power actually exerted by the steam in the cylinder without taking into account the engine friction.

By *brake horse-power* (B.H.P.), or *effective horse-power*, is meant the actual horse-power delivered by the crank shaft of the engine. Thus an engine giving 100 indicated horse-power will give only about 90 to 95 brake horse-power, the remainder being lost in friction. If the engine gives 90 brake horse-power for every 100 indicated horse-power, its mechanical efficiency is said to be 90 per cent. If it gives 95 brake horse-power for every 100 indicated horse-power its mechanical efficiency is 95 per cent.

This mechanical efficiency has nothing to do with the consumption of steam or thermal efficiency. A single-cylinder engine which is wasteful in steam has often a higher mechanical efficiency than a three-cylinder, triple-expansion, condensing engine which is very economical of steam. In comparing the merits of two engines, it is necessary to look at the consumption of steam per brake horse-power, or, if the consumption is given per indicated horse-power, it is necessary to ascertain the efficiency of such engine, and to convert the figures into consumption of steam per brake horse-power. It is important to remember that because an engine is claimed to have a very high mechanical efficiency, it does not follow that it is an economical engine to use. Before proceeding with the study of compound and triple-expansion engines, and going into the questions of economy of steam, &c., it may be well to state how the horse-power of a single-cylinder engine may be worked out.

The indicated horse-power of a double-acting engine is found by multiplying twice the stroke (in feet) by the number of revolutions per minute, by the area of the piston in inches, and by the mean pressure exerted on the piston, and dividing the result by 33,000. Stated as a formula it is expressed thus—

$$\text{I.H.P.} = \frac{2 \times S \times R \times A \times P}{33{,}000};$$

where S = stroke in feet.
R = revolutions per minute.
A = area of piston in inches.
P = mean pressure in lbs. exerted on the piston.

Let us take an actual case, say of an engine as shown by Fig. 35, the cylinder of which is 10 inches in diameter, the stroke 8 inches, the number of revolutions 300 per minute, the initial steam pressure as it enters the cylinder 75 lbs., the cut-off takes place at about ·65 of the stroke, and the average mean pressure on the piston during the whole of the stroke is 50 lbs.

As the stroke has to be worked out in feet or parts of a foot before we can make the calculation, we must find out what part of a foot the stroke of the engine—viz., 8 inches—is. If we turn to the decimal equivalents given at the end of the book, we see that 8 inches is ·666 of a foot. We must also find out what the area of a 10-inch piston is; this we see from the table of areas is 78·5 square inches.

The calculation now is

$$\frac{2 \times \cdot 666 \times 300 \times 78 \cdot 5 \times 50}{33{,}000} = 47 \cdot 5.$$

The answer is 47·5 I.H.P. If the mechanical efficiency of the engine is 90 per cent., then the brake horse-power will be 47·5 × 90 ÷ 100 = 42·75 B.H.P.

To obtain an initial pressure of 75 lbs. in the cylinder the boiler pressure would probably require to be 90 lbs. per square inch, as there is usually a drop of from 5 to 10 lbs. between the boiler and engine stop valve, the drop depending upon the length and diameter of the steam pipe, as explained in the previous chapter. There should also be a difference of 5 lbs. on the two sides of the engine governor, if good governing is desired.

Having got 75 lbs. initial pressure we cannot, however, count on this pressure all through the stroke, as after the point of cut-off, which in this case we have assumed to take place at ·65 of the stroke, the steam expands and the pressure falls. The question which will naturally be asked by a beginner at this point is—supposing the initial pressure is 75 lbs., and the cut-off ·65 of the stroke, how am I to know what the average pressure of steam will have been by the time the piston has reached the end of the stroke? Well, if the steam behaved like a perfect gas, and expanded adiabatically—*i.e.*, without receiving heat from, or giving up heat to, the cylinder walls—the answer would be simple, for, according to Boyle's law, the volume of a gas varies inversely as its pressure, the temperature being kept constant. Thus if a given quantity of gas expands to twice its volume, its pressure falls to half the original pressure; if expanded to four times its original volume, the pressure falls to one-quarter of what it was originally. This law is expressed thus, $P \times V =$ constant, for if we have gas at 100 lbs. pressure, the volume of which is, let us say, 4 cubic feet, and we multiply the pressure by the volume, the answer is 400; if we expand the same quantity of gas into a space of 8 cubic feet, its pressure will fall to 50 lbs., but multiplying the pressure and volume together, we still get 400 as the answer. Therefore, pressure multiplied by volume equals a constant. This fall of pressure is represented by a *hyperbolic curve*, which the reader will be shown how to construct.

The case of steam is not so simple as that of a perfect gas, but it follows approximately the same law.* The true adiabatic curve for steam which is initially dry, falls slightly below the hyperbolic curve, but the effect of initial condensation and re-evaporation (referred to later), causes the expansion line of diagrams taken from actual engines, to approximate closely to the hyperbolic curve. To construct this curve proceed as follows:—

Let A B (Fig. 39) represent the stroke of the engine, also the line of absolute vacuum, and A C the absolute pressure of the steam (the meaning of absolute pressure was explained on p. 47). Complete the rectangle as shown, which will then represent the volume of the cylinder. Mark D at the point of cut-off, and draw the line D E. Divide E H into equal divisions, 1, 2, 3, &c., and let fall perpendiculars to the line A B. Draw lines from the point A to the points 1, 2, 3, 4, &c., and from the points where these lines cut E D at x, x, x draw horizontals cutting the perpendiculars 1, 2, 3, &c., at the points z, z, z; then the curve E M drawn through these points is a hyperbola. The atmospheric line is drawn 14·7 lbs. above the zero line.

Fig. 39 shows how the steam pressure falls if the cut-off takes place at ·25 of the stroke. Figs. 40 and 41 show the fall of pressure when the cut-off takes place at ·5 and ·75 respectively.

In constructing the above diagrams we have neglected the effect of clearance—*i.e.*, the steam passages and space between the piston and cylinder cover when the piston is at the end of its stroke. If these clearances are to be taken into account,

---

* The expression for the relation between the pressure and volume of saturated steam is $P \times V^x =$ a constant, where the value of x depends on the dryness of the steam. The expression, $PV^{\frac{17}{16}} =$ constant, is frequently used for saturated steam.

THE STEAM ENGINE. 99

their volume must be represented by the rectangle shown by dotted lines in Fig. 41, the point A being set back accordingly.

From a hyperbolic curve it is easy to find the mean pressure theoretically exerted by the steam throughout its stroke,

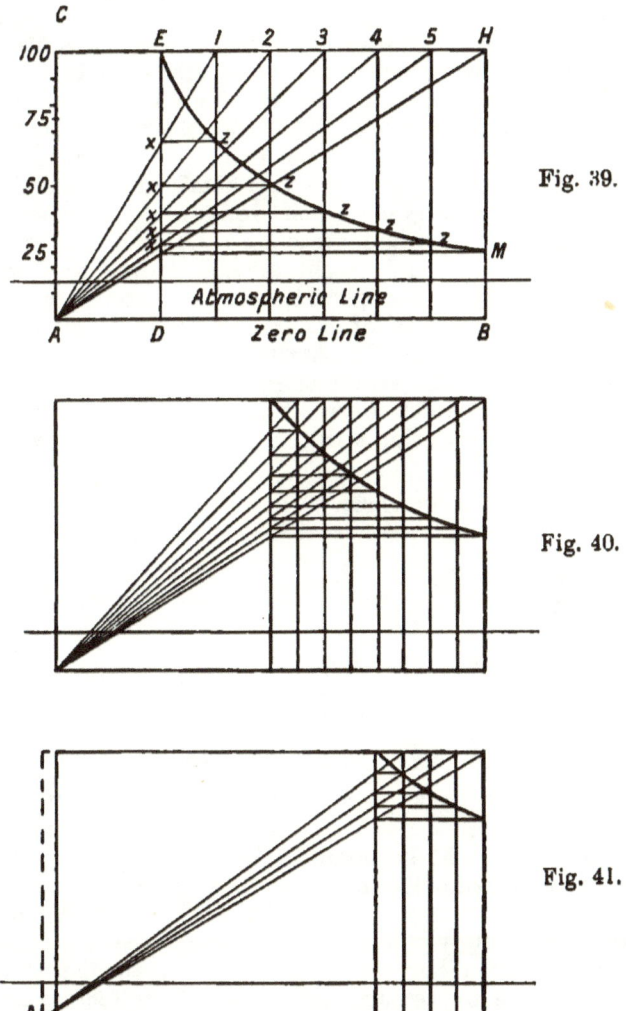

Fig. 39.

Fig. 40.

Fig. 41.

assuming $PV =$ constant. Tables giving the mean pressure, obtained with various points of cut-off, are given in many engineering pocket-books and text-books, but the mean pressures so given are, as a rule, much higher than are found in actual practice.

In the diagrams given it is assumed that the steam continues to enter the cylinder at its full initial pressure right up to the point of cut-off, but in real engines, especially in small fast-running engines, the steam gets throttled in the passages, and there is a falling-off in pressure before cut-off takes place. Again the diagrams, as shown, assume the steam to be passed away without causing any back pressure; also the effect of compression is not taken into account. In the indicator diagrams taken from an actual engine, the mean pressure which is really exerted by the steam is seen, and the data obtained from such diagrams usually serve as a guide when designing a new engine. The following Table may be instructive. Column A gives the mean pressures which should be obtained, according to a well-known engineering pocket-book, with various cut-offs, and with an initial steam pressure of 80 lbs.; column B gives the mean pressures which should be obtained, according to rules given in a good book on the steam engine; while column C gives the mean pressures which were actually obtained in a good high-speed engine, indicating about 40 indicated horse-power at full load.

INITIAL PRESSURE 80 LBS. ABOVE ATMOSPHERE.

Point of Cut-off.	A. Mean Pressure.	B. Mean Pressure.	C. Mean Pressure.
·25	47·7	32	30
·375	59·5	44	39
·5	67·7	53½	46
·625	73·5	63	51
·75	77·3	71	56

It is, of course, quite possible that the mean pressure given in column B might be obtained in a slow-speed engine having steam passages of ample size, but those given in column A could only be obtained in a theoretically perfect engine, and must be treated accordingly.

We have seen how the horse-power of an engine is worked out, if the mean pressure and all the other factors are known. If we wish to know what the mean pressure must be to give a certain power when the speed and other particulars are known, the formula is—$P = \dfrac{33{,}000 \times H.P.}{2\,SRA}$.

If the speed, mean pressure, and length of stroke are known, and we wish to find out what area the piston must have in order to give a certain horse-power, the formula is—$A = \dfrac{33{,}000 \times \text{H.P.}}{2\,\text{SRP}}$.

The formula is the same in each case, but transposed.

For working out the power of a single-acting engine—*i.e.*, one in which the steam acts on one side of the piston only—the formula is—$\text{H.P.} = \dfrac{\text{SRAP}}{33{,}000}$.

As already explained, one horse-power is the power required to raise 33,000 lbs. 1 foot high in one minute, or 1 lb. 33,000 feet high in one minute.

We have said that the single-cylinder engine is uneconomical. Why is this, and can it be made to work economically?

In the first place, with the D-form of slide valve, or with a single piston valve, the point of cut-off cannot well be made earlier than $\tfrac{2}{3}$ of the stroke, so that when the piston has completed its stroke a large quantity of steam at a fairly high pressure, and capable of doing further work, is passed away to the atmosphere, and is lost, or, in other words, so much heat is wasted (see Fig. 41). But suppose, instead of the D-valve, we take the case of an engine fitted with the Corliss or Drop type of valve, which will allow the cut-off to take place at any part of the stroke without interfering with the exhaust, and make the cut-off take place at $\tfrac{1}{8}$ or $\tfrac{1}{4}$ of the stroke, thus allowing the steam to work expansively (see Fig. 39). Would this not be economical?

It would not, for these reasons. If we cut off the admission at, say, $\tfrac{1}{8}$ of the stroke, and expand the steam until the end of the stroke, the steam, when it leaves the cylinder, is much cooler than it was when it entered; it then cools down the ports, cylinder walls, and piston, and a certain proportion of the incoming steam at the next stroke on coming in contact with the ports, cool walls, and piston, condenses and turns into water. This is called *initial condensation*, and is responsible for a considerable loss of heat or energy.*

---

* The loss of heat from the cylinder walls is due not only to the fact of their having been in contact with the cooler steam, but also to the circumstance that when the steam has condensed on the cylinder walls it remains there as a film of water. Now, after cut-off takes place, the steam expands and the pressure falls; the film of water is then in a position to evaporate. It will have been gathered from the chapter on "Steam Raising" that the temperature at which steam is formed depends on the pressure. Now, the pressure having fallen, the water robs the walls of the heat which had previously been given to them, and re-evaporates. This loss has to be made good by the incoming steam at the next stroke.

From data in his possession, the author finds that a good single-cylinder non-condensing engine of the high-speed type, working with 75 lbs. initial gauge pressure and ·7 cut-off, used $31\frac{1}{2}$ lbs. of steam per I.H.P. per hour; while with the same pressure and an extremely early cut-off—viz., ·2—the engine used 32 lbs., so that the early cut-off was actually harmful. The best point of cut-off for this engine was ·5, when it used $29\frac{1}{2}$ lbs. of steam per I.H.P. per hour. It should be remembered, too, that a good high-speed engine suffers less from initial condensation than a slow-speed engine, as in the latter there is more time for the interchange of heat to take place between the periods of admission.

This initial condensation prevents one from obtaining much advantage from taking the exhaust steam from a single-cylinder engine to a condenser, as the ports and cylinder walls are subjected to a still lower temperature than when non-condensing.

Cushioning in a single-cylinder non-condensing engine, and in the H.P. cylinder of a compound condensing engine has a beneficial effect on the economy of the engine, as not only is a certain volume of steam saved, but also the temperature of the compressed steam is raised, and the initial condensation of the incoming steam is reduced.

To use steam to its greatest advantage we need to have it at as high a temperature as possible (within limits), to make it do as much useful work as possible, during which time it will be giving up its heat, and then to get rid of it without cooling down the incoming steam more than is necessary. The greatest ideal efficiency of a steam engine is

$$\frac{T1 - T2}{T1};$$

where $T1$ = temperature of steam supplied.
$T2$ =   „   „   rejected.

The conditions just mentioned can be more nearly complied with in a steam turbine than in a reciprocating engine. To approach them in the latter we need two, three, or even four cylinders.

A compound engine is one in which the expansion of the steam is carried out in a pair, or pairs of cylinders. In a two-cylinder compound engine there is one high-pressure (H.P.) cylinder and one low-pressure (L.P.) cylinder. The steam, after doing its work in the H.P. cylinder, instead of being passed away to the atmosphere or condenser, is taken to a receptacle or receiver, and from this receiver the L.P. cylinder draws its steam. The L.P. cylinder is much larger in diameter than the H.P. cylinder, the area of the L.P. piston being usually three or three and a half times greater than that of the H.P. piston. As soon as the valve of the L.P. cylinder opens, steam enters from the receiver and continues to do so until the cut-off takes place, when the steam expands and forces the piston to the end of its stroke; the exhaust port then opens, and the steam is either passed to the atmosphere or to the condenser. Thus the H.P. cylinder is never cooled down by the comparatively low final temperature of the out-going steam.

When the steam is exhausted to the open air it has to be discharged against the pressure of the atmosphere, which is about 14·7 lbs. (temperature of the steam, 212° F.). If, however, the steam is condensed by means of cold water, a vacuum is formed, and the steam is discharged against an absolute pressure of 1 or 2 lbs. only (the exact pressure depends on the condensing arrangements and size of the exhaust pipe), and at a temperature of 100° to 125°. Thus, more work is got out of the steam than can be obtained when discharging against atmospheric pressure.

In the earliest form of steam engine steam was used for raising the piston only. When the piston reached the top of its stroke a jet of water was squirted in, a vacuum was created, and the atmospheric pressure forced the piston down. The initial condensation must, of course, have been enormous.

In a triple-expansion engine the expansion of the steam is carried out in three cylinders; in such an engine the range of temperature in each cylinder is still less than in a compound engine, and the loss from initial condensation is therefore still further reduced.

The consumption of steam, not superheated, in a good compound engine of the Corliss type, when condensing, is between 14 and 15 lbs. of steam per I.H.P. per hour. When exhausting to the atmosphere the consumption is between 19 and 22 lbs. per I.H.P. per hour.

The consumption of steam in a good triple-expansion engine when condensing is between $12\frac{1}{2}$ and $13\frac{1}{2}$ lbs. per I.H.P. per hour. When exhausting to the atmosphere the consumption is between $18\frac{1}{2}$ and $19\frac{1}{2}$ lbs. per I.H.P. per hour, assuming the steam pressure is not less than 180 lbs. above the atmosphere. With steam pressures lower than this there is but little advantage in employing a triple-expansion engine, when it has to work non-condensing.

The consumption of steam in a quadruple-expansion engine is slightly less per I.H.P. than in a triple-expansion engine, but it is somewhat doubtful whether the gain through carrying out the expansion in four stages instead of three is sufficiently great to compensate for the extra first cost, and for the extra friction due to the use of the fourth cylinder.

In marine engines, which are usually tested for economy after erection on board ship, the economy or otherwise is usually referred to in lbs. of coal, as it is rather a troublesome matter to weigh or measure the water used. The average consumption of coal on modern battle-ships having triple-expansion engines is 1·78 lbs. of coal per I.H.P. per hour at the most economical load —viz., about 70 per cent. of the full power, and 1·92 lbs. of coal at full power.

In the *Britannia*, where superheated steam was used, the consumption of coal was 1·5 and 1·85 lbs. respectively.

These consumptions of coal include the amount burnt for making steam for the auxiliary machinery. In the mercantile marine, where there is less auxiliary machinery, the consumption of coal is from 1·3 to 1·75 lbs. per I.H.P. per hour.

The consumption of coal in locomotive engines, as tested in America by means of a dynamometer, varied from 3·5 to 4·5 lbs. per B.H.P. in simple engines for goods traffic, and from 2 to 3·7 lbs. in compound engines for the same class of work. The consumption of coal in simple passenger locomotives varied from 3·5 to 5·0 lbs., and in compound passenger locomotives from 2·2 to 5·0 lbs., the higher consumption of coal always occurring at high speeds.

In a mill having a good compound condensing engine a consumption of 1·5 to 1·75 lbs. of coal per I.H.P. is regarded as a very fair performance.

Corliss Gear.—We have already spoken of an engine fitted with the Corliss type of valve gear, by means of which the cut-off can be made early without interfering with the opening of the exhaust port at the proper time. Fig. 42 shows the Spencer-Inglis form of this gear in elevation, and Fig. 43 shows a section through the L.P. cylinder of a large Corliss engine recently supplied by Messrs. Fullerton, Hodgart & Barclay, of Paisley, for the East Randt Gold Mining Company.

The Corliss form of valve gear was invented by an American engineer, whose name it bears. It has been very largely used in England and America, but is only suitable for engines running at a speed of 120 revolutions per minute or less.

It will be seen from Figs. 42 and 43 that there are separate admission and exhaust valves at each end of the cylinder; by this arrangement long steam ports are avoided, and the relatively cool exhaust steam does not pass through the admission ports, as is the case with an engine having a single **D**-slide valve. The exhaust ports, which are at the lowest portion of the cylinder, allow any water which has not re-evaporated to be swept out by the piston at every stroke.

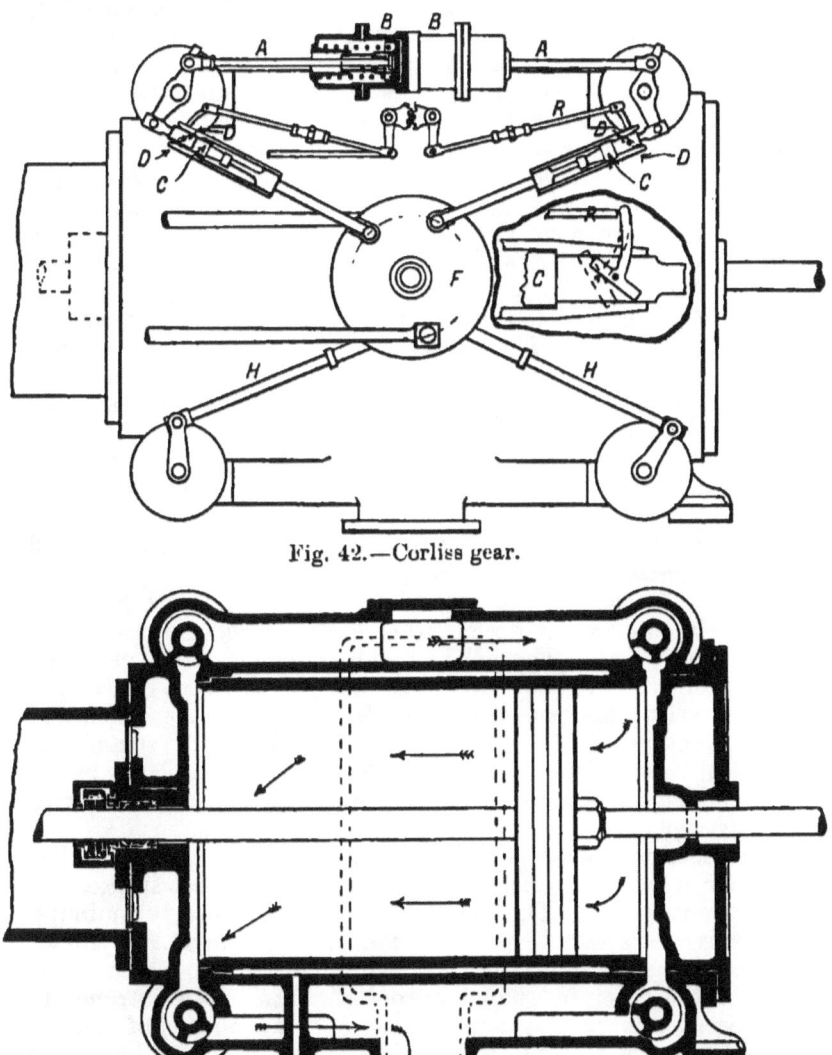

Fig. 42.—Corliss gear.

Fig. 43.—L.P. cylinder, with double-ported Corliss valves.

The Corliss gear may perhaps look complicated to the beginner, but really it is very simple. The steam-admission valves at the top of the cylinder are kept closed by the rods A A, Fig. 42. Each of the rods is connected to a plunger inside the small cylinders and dashpots placed back to back at B B. The plungers usually are drawn in by means of springs, and, less frequently, by placing one side of the plunger in communication with the condenser, and leaving the other side open to the atmosphere. The effect is to keep the main steam-admission valves closed until one of the blocks, C, which is connected to a lever fixed to the valve spindle, is taken hold of by the catch plates D D; the motion of these catch plates and the block they have caught hold of causes the steam valve to open. When the piston has reached a certain position the catch plates are opened outwards, the block C is released, and the main steam valve immediately closes.

The mechanism which causes the plates to open outwards is shown by the inset. The releasing toe is pivoted to a prolongation of the block C already referred to, and when drawn along with it—the end of the lever attached to the toe being prevented from moving inwards by the rod R—the toe is forced to take up a more oblique position across the catch plates, and opens them outwards. The rod R, is also connected to the governor, so that if the engine runs too fast the toe is forced into its oblique position a little earlier in the stroke; the point of cut-off is thus regulated by the governor.

The catch plates D are worked from the wrist-plate F, which receives its oscillating motion from an eccentric, or eccentrics, in the ordinary way. The exhaust valves have no trip gear like the steam valves; they are connected by the rods H H to a second wrist-plate placed behind the first, and are unaffected by the action of the governor. Sometimes both steam and exhaust valves are worked off the same wrist-plate, but a greater range and more accurate setting can be obtained with two wrist-plates.

The sectional view (Fig. 43) is almost self-explanator. The cylinder shown is 40 inches in diameter, has a 5 feet stroke, and is steam-jacketed. The speed is 51 revolutions per minute. The valves are double ported. Steam from the H.P. cylinder enters at the far side, and passes up the passage, which forms part of the receiver, shown by dotted lines. The entrance to this passage is beyond the exhaust opening, and is, of course, separated from it. The steam valve at the top right-hand end has been drawn open by the catch plates, while the steam valve at the top left-hand end, which had previously been tripped, is closed. The exhaust valve at the bottom right-hand end is closed, while the exhaust valve at the other end is open. The opening K is for the purpose of draining the jacket. The admission to the jacket is at the far side of the cylinder, and is not shown. The glands of the piston and tail rods are packed with a patent metallic packing, which is described later.

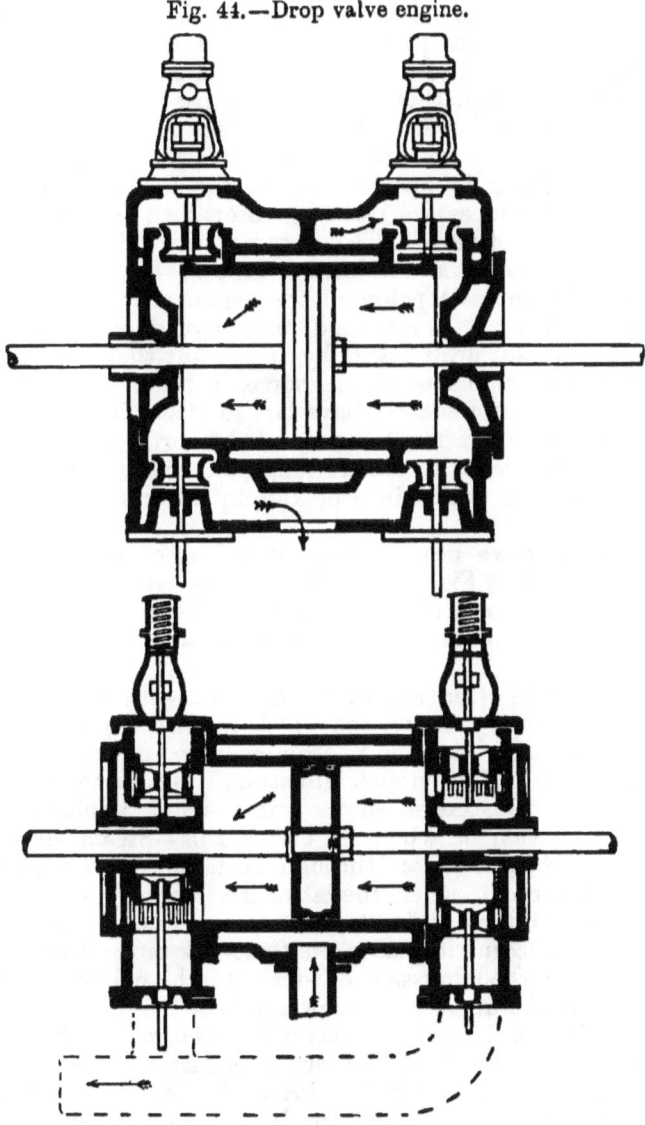

Fig. 44.—Drop valve engine.

Fig. 44a.—Van der Kerchove engine.

Drop, or Double-beat Valves.—Another form of valve, which is considered by some engineers to be more suitable for use with superheated steam than the Corliss valve, is the double-beat valve. The L.P. cylinder of an engine fitted with these valves is shown by Fig. 44. The valves, which are of the equilibrium type, are usually worked by some form of trip gear actuated by eccentrics placed on a shaft running alongside the cylinder, and parallel to the piston-rod. A considerable number of engines fitted with this form of valve has been made in this country and abroad. The chief objection to them is that the valves close with a somewhat heavy blow, causing wear and tear, and resulting in leaky valve seatings. Another objection is that the clearances are rather greater than is the case with the Corliss form of valve. To overcome these objections Messrs. Van der Kerchove, a well-known firm of Belgian engineers, have brought out and patented the engine shown by Fig. 44a, in which piston valves are used instead of double-beat valves. The piston valves are worked by trip gear; they are provided with piston rings and work, of course, in liners in which ports are cut. The piston valves have lap, so that the cushioning of the valve is effected after it has closed the port. The clearances, as will be seen from the illustration, are extremely small. This type of engine is made in England by Messrs. Musgrave, of Bolton.

The Lentz valve gear consists of valves of the double-beat type, shown by Fig. 44. The valves are, however, not worked by any form of trip gear, but take their motion from a rocking cam actuated by eccentrics, and the usual link-reversing gear.

The Willans Central-valve Engine.—An engine of an entirely different type, and one which has had a great vogue in electric generating stations in this country, is the high-speed Willans central-valve engine, illustrated diagrammatically by Fig. 45. The illustration shows a three-crank triple-expansion engine, the action of which is briefly this—Steam enters the steam chest A, and passes through ports cut in the trunk or hollow piston-rod T, when the latter is at the top of its stroke. Moving up and down inside each trunk is a line of piston valves worked by an eccentric placed on the crank pin. These piston valves control the admission of steam to the cylinder, and the exhaust from the upper to the lower side of the piston, the space below each piston being the receiver R. The steam, after having done its work in the H.P. cylinder B, is passed to the underside of the piston or receiver R. From this receiver the intermediate cylinder D draws its steam, and in turn passes it on to its receiver R_1, and thence to the L.P. cylinder E; after leaving the L.P. cylinder, the steam enters the exhaust chamber G, which is placed in communication either with the atmosphere or with the condenser.

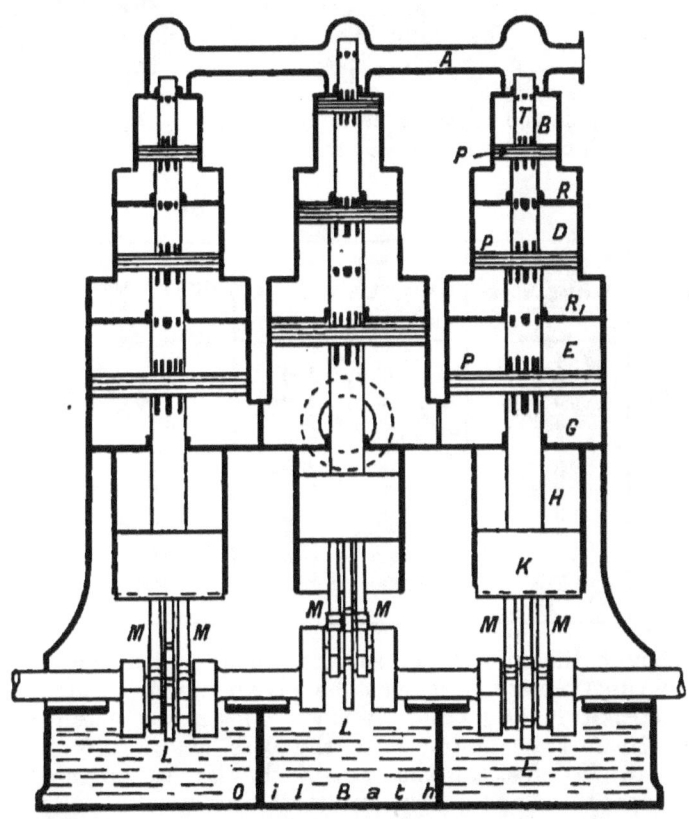

Fig. 45.—Triple-expansion three-crank Willans engine.

A, Steam chest.
B, H.P. cylinder.
R, H.P. receiver.
D, Intermediate cylinder.
R_1, Intermediate receiver.
E, L.P. cylinder.
G, Exhaust chamber.
H, Air chamber.
K, Guide piston.
T, Trunks.
L, Eccentrics and straps.
M, Connecting-rods.
P, Piston.

The Willans engine is single-acting—*i.e.*, the steam acts on one side of the piston only, so that there is no push-and-pull action on the brasses. In order to prevent a pull coming on the brasses on the up-stroke due to the momentum, or more scientifically expressed, to the *inertia* * of the moving parts, a separate chamber H is provided, in which air is compressed during the upward stroke, the greater portion of the power expended in compressing the air being given out on the downward stroke. The piston K, which compresses the air, serves also as the cross-head and guide. The crank chamber below the crank shaft is partly filled with oil, which is splashed up over the journals and into the guide cylinders.

The horse-power of some standard Willans engines and the speeds at which they run are as follows:—

TABLE XV.

Makers' Distinguishing Letter.	I.H.P.	Revolutions per Minute.
3 F	90 to 100	470
3 G	120 ,, 150	460
3 H	200 ,, 240	380
3 I	300 ,, 350	350
3 Q	400 ,, 450	340
3 R	500 ,, 575	320
3 S	600 ,, 700	300
3 T	750 ,, 825	270
3 V	1,000 ,, 1,250	230
3 X	2,000 ,, 2,500	200

Although the number of revolutions at which the Willans engine runs is very high, yet the piston speed, owing to the short stroke, is comparatively low. The piston speed ranges from 470 feet per minute in the case of the 100 I.H.P. engine to 787 feet in the case of the largest engine of all. These speeds compare favourably with those of many large marine engines and locomotives; for instance, the piston speed of the engines of H.M.S. *Africa* is 1,124 feet per minute when running at full speed—viz., 128 revolutions per minute; and 920 feet when running at cruising speed—viz., 115 revolutions. The piston speed in a modern locomotive having cylinders 19¼ inches diameter by 26 inches stroke and 6-feet 6-inch driving-wheels, is 1120 feet when running at 60 miles per hour. The

* It should be explained that a body requires a certain force to set it in motion; when once set in motion the body will continue to move in a straight line until stopped by gravity or by some other force. This unwillingness to start and unwillingness to stop when once set going is called the *inertia* of the body.

piston speeds of large horizontal Corliss engines are usually about 500 feet per minute.

Belliss and Browett-Lindley Engines.—Two other well-known high-speed engines are those of Messrs. Belliss and Browett-Lindley. These engines are double-acting, the principle being the same as that of the engine shown by Fig. 35. The engines, however, are usually made either compound or triple-expansion. The compound engine has one high-pressure and one low-pressure cylinder placed side by side and two cranks; the triple engine has three cylinders placed side by side and three cranks. It may be here remarked that a three-crank engine, especially if the pistons are approximately of the same weight, causes much less vibration than any two-crank engine; for this reason, three-crank engines are usually selected for those electric generating stations which are surrounded by dwelling-houses. The vibrations from two-crank high-speed engines, even if mounted on heavy foundations, are sometimes transmitted to considerable distances, especially if the soil is moist, and may cause considerable annoyance to residents during the night time. The vibrations set up by a slow-speed engine, as shown by a vibration recording instrument, may be greater than those of a high-speed engine, but the period of vibration being slower the annoyance caused is not so great.

Both the Belliss and Browett engines are forced lubrication engines—*i.e.*, the lubricant is forced into the bearings under pressure, a pump driven off the crank shaft being provided for the purpose. The powers and speeds of these engines are approximately the same as those of the Willans engine already given. It is owing to the short stroke of these engines and to the system of forced lubrication that they are able to run at such high rotative speeds.

A few words as to the merits of each type of engine, and as to the disadvantages connected with its use, may now be said.

Slow-speed Horizontal Engines.—These engines are very largely used in mills and works of all kinds where fair economy, freedom from trouble and breakdown, and a long life are the most important considerations. The engines usually found in mills are of the compound type. A compound engine is not quite so economical as one of the triple-expansion form, but the former works with very fair economy. The consumption of dry, but not superheated, steam when condensing is about $14\frac{1}{2}$ lbs. per I.H.P. per hour. The wear and tear of such engines, if well made and properly lubricated, is very small, and the engine requires but little attention.

The fact of the heavy pistons rubbing on the lower half of the cylinder is not a good feature, and for many years past it has been the custom to prolong the piston-rod, so as to pass through the back cover, and provide it with a block running on a slide, the idea being to carry the weight of the piston. A circular rod, however, is almost the worst form that a beam designed to carry a weight can take, and it is now considered by many engineers that it is better to make the piston of great width, so as to provide ample bearing surface, and to dispense with the tail rod. It is claimed that the wear of a cylinder having a wide and light piston is more even, and less in extent, than is the case with a cylinder having a narrow piston supported by a tail-rod which sags. In any case a cylinder, the liner of which is made of good hard cast iron, and has been truly bored in the first place, will last for a great many years without requiring to be re-bored.

The reason why horizontal engines are not employed in large electric generating stations is because they cannot run at speeds sufficiently high to enable them to be directly coupled to the dynamos, and the space required for driving the latter by ropes cannot be given. In any case the use of ropes, involving a loss of 5 per cent. or 7 per cent. would not be tolerated.

Vertical Slow-speed Engines.—These are employed to the exclusion of almost all other types of reciprocating engine for marine work. They occupy less floor space than a horizontal engine, but are more expensive to construct. The frame for carrying the cylinders requires to be much heavier and stronger than is necessary in the case of a horizontal engine, the cylinders of which in land work rest on concrete. Platforms and ladders, too, are required with vertical engines, thus adding to the expense.

A good feature about vertical engines is that there is not the same risk of the cylinders wearing oval as in a horizontal engine, as the piston does not bear upon the cylinder walls. A vertical engine lends itself more readily to the triple-expansion form than a horizontal engine, and in some mills and works where first cost is of less importance than extreme economy of steam, vertical triple-expansion engines may be found. Vertical engines of the marine type are used in some electric generating stations, but their comparatively slow speed necessitates a very large and costly generator if the latter is to be coupled direct to the engine.

High-speed Single-acting Engines.—The Willans central-valve single-acting engine was introduced at the time when electric lighting was first being carried out upon a considerable scale, and when it was found necessary to install as much power as possible in a small space. The high speed of the Willans engine enabled it to be coupled direct to the dynamo; thus the space which would otherwise have been required for a belt or rope drive was saved, also the loss due to this method of transmission was avoided. In addition, the engine was remarkably economical in steam, and was capable of making very long runs without a stop. It is on record that one of these engines ran for seven months night and day in a copper depositing works without a single stop. The success of the Willans engine in important generating stations, such as those of the St. James and Pall Mall Company, and of the Westminster Company, led to its being widely adopted all over the country, even in places where ample space was available, and for many years this engine practically held the field for electric lighting and traction work. In addition to the advantages of high speed and economy, one other must be referred to. The engine is single-acting, and, owing to the action of the air buffers (already referred to), the constant thrust principle is really carried out. As there is no push-and-pull action, wear of the brasses does not give rise to a knock, and, within limits, is unimportant. The engine, therefore, does not require frequently to be laid off, and a mechanic's time occupied in setting up and adjusting brasses.

The Willans engine has been employed in a few cases for driving cotton and flax mills, for which purpose the large number of impulses per minute and consequent even turning render it very suitable. It is stated that the output of a flax mill in Ireland was increased to the extent of 5 per cent. through freedom from breakages of thread, by the substitution of a Willans three-crank engine for one of the slow-speed type.

The disadvantages of a single-acting engine are as follows:—
The piston must have an area twice as great as that required in a double-acting engine of equal power, running at the same speed. If a knock is heard in a single-acting engine, such as the "Willans," it is advisable to take the engine down at once to ascertain the cause, for in the Willans engine the pistons are mounted on cast-iron trunks, through which ports are cut, and through which steam is admitted to the cylinders. If a broken piston ring or valve ring should get into one of these ports while the engine is running, it might cause a very serious breakdown. As a fact, however, breakdowns of this nature are comparatively rare.

In the Willans engine, too, there is a constant, though small, loss due to compressing and expanding the air in the air chambers. This loss prevents quite such a high mechanical efficiency being obtained, as in a double-acting engine, in which there are no air buffers.*

Double-acting High-speed Engines.—The success of the Willans single-acting engine incited makers of double-acting engines to construct them to run at speeds approximating to those of the former, and by extremely good workmanship and by the use of forced lubrication, they succeeded. The best known high-speed double-acting engines are those made by Messrs. Belliss and by Messrs. Browett & Lindley.

The advantages of these engines for driving dynamos are those possessed by the Willans engine, and the engines, conforming more nearly as they do to the ordinary type of slow speed engines, can be overhauled by men who have had no special training. The solid steel piston-rod in these engines is preferred by many to the cast-iron trunks in the Willans engine.

For electric generating purposes a high-speed engine has in the past been a necessity, but its place is seriously threatened by the advent of the steam turbine. For mill work, and in cases where high speed is not a necessity, engines of the long-stroke slow-running type are likely to hold their own against their lighter and quicker running rivals.

Steam Jacketing.—The practical advantages or otherwise of jacketing steam cylinders—*i.e.*, keeping them surrounded by live steam—has been under discussion for a great many years past. It is claimed that by jacketing a cylinder and imparting heat to its walls while expansion is taking place, and during exhaust, the fall in temperature of the walls is reduced and initial condensation is lessened. On the other hand, a certain proportion of the steam in the jacket is condensed, and opponents of jacketing say that as much steam is lost in the jacket as is saved in the cylinder.

Experiments appear to show that there is a gain through

---

* If this little book should fall into the hands of a station engineer, who would like to ascertain the loss of efficiency due to the air buffers, and who has been unable to ascertain the loss by means of an indicator card, he can easily do so, assuming the engine is coupled to a direct current dynamo, by the following method :—The first time the engine is down for any purpose, remove the steam cylinders, pistons, and high-pressure trunks, leaving the guide piston, low-pressure trunks, and air buffer covers in position. Run the dynamo as a motor, and note the current required to drive the crank shaft at its full speed, with the guide pistons and low-pressure trunk running—*i.e.*, with the buffers in action. Then remove the guide pistons, and note what current is required to run the crank shaft at full speed—*i.e.*, with the buffers out of action. The difference of current (neglecting the small difference of dynamo efficiency) will show the loss due to compressing the air in the buffers. The loss is not truly shown by an indicator card.

jacketing, if properly carried out. This gain is probably due to the fact that it is better to condense the steam outside the cylinder than inside it, as it is the film of water inside the cylinder walls which enables the transfer of heat to take place so easily, and which robs them of so much heat when re-evaporation takes place. Whether the gain in actual practice is as great as in experimental trials is, however, open to doubt, for it must not be overlooked that for the jacket to be effective the condensed steam must be got rid of, and those who have had some experience in connection with steam traps know that they have a tendency either to stick, or to leak. If the trap sticks the advantage of the jacket is lost, and if the trap leaks it is quite possible that the steam lost in this way may equal that saved in the cylinder.

There is, however, one advantage in jacketing an engine which the maker cannot afford to overlook—it makes a good selling point. If A's representative can say to an intending purchaser, "My engine is jacketed, while B's is not," it may possibly turn the scale in A's favour; and it must be remembered that, distasteful as the fact may appear, the majority of engine builders are in business, not for the pleasure of the thing, nor for the purpose of contributing papers to learned societies, but to make money for themselves, or for their shareholders.

Calculating the Power of Compound Engines.—The reader has seen how to work out the power of a single cylinder, or simple engine; to calculate the horse-power of a compound or triple-expansion engine, the formula for the simple engine holds good; but if the student has not got the actual indicator diagrams before him, care must be taken to estimate the mean pressures rightly. With condensing engines it is also more convenient to make the calculation with absolute pressures—$i.e.$, including the atmosphere. Instead of saying 75 lbs. above the atmosphere we should say 89·7 lbs. absolute, or, roughly, 90 lbs. absolute.

We will now take the case of a double-acting compound condensing engine, having one H.P. cylinder 10 inches diameter and one L.P. cylinder 18 inches diameter, the stroke of each 8 inches. The speed, 300 revolutions per minute. In working out the power of the simple engine with approximately 90 lbs. absolute pressure, we assumed the mean pressure to be about 65 lbs. absolute. If, however, instead of passing the exhaust steam direct to the atmosphere, we take it to another and larger cylinder to do more work, the pressure of the steam acting on the large L.P. piston will also exert a back pressure upon the small H.P. piston, so that, instead of having a mean effective pressure of 65 lbs. absolute on the H.P. piston, we shall only have an effective mean pressure of about 40 lbs. The mean pressure in the L.P. cylinder will, of course, be much less, as its area is about $3\frac{1}{4}$ times greater than that of the H.P. cylinder, and it only receives its steam after this leaves the H.P. cylinder. If in this case we cut off at ·65 of the stroke, we shall expand the steam about 5 times altogether. The effective mean pressure we might get in the L.P. cylinder would be about 20 lbs. absolute. We should, therefore, have to take the area of the 10-inch piston—viz., 78·5 inches—as having an effective pressure of 40 lbs., and the area of the 18-inch piston—viz., 254·4 inches—as having an effective pressure of 20 lbs. The result would be—78·5 × 40 = 3,140 lbs., and 254·4 × 20 = 5,088 lbs., or 8,228 lbs. total effective pressure on the two pistons. For the sake of simplicity in the calculations, it is usual to refer this total pressure to a corresponding pressure on the L.P. piston. In this case the area of the L.P. piston is 254·4 inches; and if we divide the total effective pressure on the two pistons by this area, we shall have the corresponding effective pressure on the L.P. piston. Thus, 8,228 ÷ 254·4 = 32·3 lbs. mean pressure, so that in our calculations for horse-power we may ignore the H.P. piston, and take into account the L.P. piston only, with the corrected mean pressure referred to it.

The formula for horse-power, as stated before, is—$\dfrac{2SRAP}{33,000}$.

The calculation for the compound engine will, therefore, be—
$$\dfrac{2 \times ·666 \times 300 \times 254·4 \times 32·3}{33,000} = 99·5 \text{ I.H.P.}$$

Testing Steam Engines.—There are two methods of ascertaining how much steam an engine uses. The first, by weighing the water pumped into the boiler, and the second, by weighing the steam (condensed) as it leaves the engine. The first method is usually adopted when it is desired to make a test after an engine has been fixed at the site and is doing actual work; but, unless precautions are taken, this method frequently makes the engine appear to use more steam than is really the case. For instance, should there be a small leak in the furnace of the boiler, any water which escapes is immediately turned into steam, and probably passes away with the furnace gases unnoticed.

To avoid such a loss, it is advisable to make a "still" test of the boiler over a period of 10 or 12 hours. For this test, steam is raised in the boiler to within 5 or 6 lbs. of the pressure at which the safety valves blow off. The level of the water in the boiler is carefully observed, all valves are then closed (the steam pipe should be closed with a blank flange), and steam maintained at a pressure below blow-off pressure. At the end of 10 or 12 hours, water which has been weighed is pumped into the boiler until the original level is reached. This weight of water is the amount of boiler leakage during the 10 or 12 hours' test. The leakage is often considerable, even in a boiler which has appeared to be tight when tested under hydraulic pressure.

In carrying out an engine trial by this method, care must be taken to see that during the trial the boiler safety valve does not blow off, that there are no leaks of steam from the steam pipe connecting the boiler to the engine, and that no auxiliary machinery, such as a separate donkey feed pump, is supplied from the boiler into which water is being weighed.

The second method is more accurate so far as the engine is concerned, and is that usually adopted at the works of engine-makers who are in the habit of testing their engines before delivery. By this method the steam as it leaves the engine is condensed and passed into a tank, which is placed on a weighing machine. It is certain that all steam which goes into an engine must come out of it, otherwise the engine would become choked with water.

If the engine is ultimately to work non-condensing, and non-condensing results are required, the chamber in which the steam is condensed is left open to the atmosphere, so that there is no vacuum. By this system of measurement boiler leaks are immaterial, as the engine is only debited with the exhaust steam which comes out of it. The boiler, too, may drive any auxiliary machinery while the test is being made.

A test of one or two hours' duration made in this way is more accurate than a 10 or 12 hours' test made by the method first described, as readings can be taken every few minutes, and if the readings tally with one another, the result may be relied upon implicitly.

To take an actual case. Suppose an engine which is giving 100 horse-power passes regularly every minute 25 lbs. of water into the tank, we know that the engine is using 1,500 lbs. of steam per hour. If now we divide this total water by the horse-power, we see that the engine is using 15 lbs. of steam per horse-power per hour. It is not necessary to have a tank of very large proportions, for, if the readings tally, it does not matter how often the tank is emptied. The arm of the weigh-bridge upon which the tank is placed is usually arranged so that as soon as a given quantity of water has passed from the engine into the tank the arm rises, the observer notes the time and pushes forward a weight which depresses the arm; as soon as the given quantity of water has again entered the tank, the arm again rises and the observer again notes the time, pushes forward the weight, and so on.

Having got the boiler, engine, and weighing apparatus, how is a load to be provided for the engine, assuming we wish to test it before it leaves the maker's works?

If the engine is coupled direct to a dynamo the matter is comparatively easy; all that is necessary is to provide resistances in the form of wires through which the current will have to travel; the power of the engine will be absorbed in generating the electric current, and the current will be dissipated in the form of heat in passing through the wires of low conductivity. If suitable wire resistances are not available, a resistance can be formed by immersing two metal plates in a tank of water, and making the current force its way through the water between them. Such a water resistance is, however, not so satisfactory as a wire resistance, as it is difficult to keep the load steady. The electrical method of providing a load for the engine is a good one, as if the volts and amperes given by the dynamo are noted and its efficiency is known, the actual brake horse-power given by the engine can be calculated. Chapter xi. will show how such calculations are made. If, however, the engine is not coupled to a dynamo, a load can be put on the engine by means of a brake.

This brake, too, can be made to show how much horse-power the engine is giving. In this way; we know that 1 horse-power is the equivalent of lifting 33,000 lbs. 1 foot high in one minute. Now, if we were to arrange matters so that the friction caused by a band surrounding a pulley which was travelling at the rate of 1 foot in one minute, exerted a pull on the band of 33,000 lbs., this would be equivalent to lifting 33,000 lbs. 1 foot in one minute. A rope-brake working on this principle is frequently used. It consists of several pieces of rope arranged to form a flat band (similar to the plaited leather horse-girths sometimes used when schooling young horses). The rope is well greased, and is lapped round the pulley of the engine to be tested, one end is attached to a portable weighing machine of the "Denison" type, preferably carried by a crane, the other is attached to a strong spring balance. Of course, in actual practice the face of the pulley travels much faster than 1 foot in one minute, and the pull on the ropes is much less than 33,000 lbs. for every horse-power, but these figures make the principle clear.

In the case of a brake load, the work done by the engine is absorbed in friction, and passes away in the form of heat. The fact that the work done by the engine passes away as heat, does not seem to be clearly realised by all of those who have to put a load on an engine by means of a brake. The author knew one individual who felt distinctly aggrieved because the ropes forming the brake got too hot and charred, in spite of his having used tallow, &c. He complained that he had to throw buckets of water on the brake ropes and pulley to keep them cool, thus causing a great mess on the floor.

In case the reader would like to see approximately how much water would be required to be thrown on a rope-brake absorbing 100 H.P., in order to keep it cool, neglecting the dissipation of heat by radiation and conduction, the method of making the calculation is explained. It is as follows:—100 H.P. × 33,000 ft.-lbs. = 3,300,000 ft.-lbs. of energy to be got rid of in one minute. We read in Chapter III. that 772 ft.-lbs. are the equivalent of 1 B.T.U., also that 1 B.T.U. will raise 1 lb. of water by 1° F., so that, if we divide 3,300,000 by 772, we shall see how many thermal units have to be got rid of, and how much water will be required. 3,300,000 ÷ 772 = 4,274 B.T.U. Assuming the water in the buckets is at a temperature of 60° F., and after being thrown on the brake its temperature is raised to 212°F., each lb. of water will take away 152 thermal units, and a gallon of water (weighing 10 lbs.) will take away 1,520 B.T.U. Now, if we divide 4,274 by 1,520, the answer is 2·81 gallons of water per minute, so that 2·81 gallons of water per minute will require to be thrown on the brake to keep it cool when absorbing 100 H.P. If, however, instead of throwing water on the brake by the bucketful, and only raising its temperature to 210° or 212° F., matters could have been arranged so that the water was thrown on the brake in a fine spray, say from a hose with a fine rose, and the whole of the water was turned into steam, a very much smaller quantity of water would have been required, for we read in Chapter III. that a very much larger quantity of heat is required to evaporate a lb. of water at 212° than is required merely to raise it to this temperature. Let us see how much water would be required to keep the brake cool if all the water thrown on to it were vaporised. It has already been said that the latent heat of 1 lb. of steam at 212° F. is 966 B.T.U., therefore the water, if turned into steam, will absorb all this heat from the brake, as well as the 152 B.T.U. required to raise the water from 60° to 212°. Every pound of water thrown on the brake will, therefore, absorb 966 + 152, or, say, 1,118 B.T.U., and as we have seen that 4,274 B.T.U. must be got rid of per minute, 3·82 lbs., or a little more than ⅓ of a gallon of water per minute, will answer the purpose.

A sound and working knowledge of the principles of heat and steam would have shown the engineer, to whom we have referred, (1) that the brake was bound to get hot when absorbing power; and (2) that by using smaller quantities of water, and turning most, if not all, of it into steam, he could have prevented a good deal of the mess on the floor.

It will be realised, from the foregoing, that any form of friction brake, unless suitable means are provided for taking away the heat, is quite unsuitable for absorbing large powers. When dealing with engines of small power, sufficient heat is lost by radiation to enable the brake to keep fairly cool.

A much more convenient form of brake is the Froude water brake (constructed by Mather & Platt, of Manchester). In this brake there is an internal wheel, carried on a long shaft, having a coupling at one or both ends; the wheel consists of two saucer-like vessels placed back to back, each divided by vanes having a forward rake. The external casing, which is free to turn on the shaft, has two similar saucer-like vessels, also divided by vanes having a backward rake. This external casing carries a long arm or lever, at the end of which weights are suspended. The engine which is to be tested is coupled to the flange on the shaft carrying the wheel, and water is admitted to the brake. The rotation of the wheel acting on the water tends to make the casing rotate, but it is prevented from doing so by the weights carried at the end of the arm. The rotary action, however, causes the arm to rise and lift the weights, and the weights being known, the actual power can be calculated.

The water passes through the brake in a continuous stream, the work done being in the form of heat imparted to the water

Such a brake is somewhat expensive, especially if of large size, and is only possessed by a few engineering firms, but its value may be realised from the following incident which came to the author's knowledge a short time ago. A firm of engine makers accepted a contract for some engines, which were guaranteed to give a certain mechanical efficiency. The engines were duly constructed, and coupled to some dynamos, which were also guaranteed to give a certain mechanical efficiency. When the plant was tested, the combined efficiency was not equal to that which would have been obtained had the engine maker's and dynamo maker's guarantees been kept. Perhaps we had better explain what is meant by combined efficiency. Let us say that the engine maker guaranteed a mechanical efficiency of 90 per cent. and the dynamo maker an efficiency of 90 per cent., the combined efficiency of the plant should be 81 per cent. That is to say, the actual electric output should be 81 H.P. for every 100 H.P. developed in the engine cylinders.

Fig. 46.—Crosby indicator.

Fig. 46a.—Crosby indicator for superheated steam.

The efficiency in the case in question not being obtained, the question arose as to who was in fault. The power of the engines being too great to admit of the use of a rope brake, the matter was settled by the purchase of a Froude brake, when it was found that the fault lay with the dynamos.

We will assume that all arrangements have now been made for giving the engine a suitable load; how are we to ascertain exactly what I.H.P. the engine under test is giving? The stroke of the engine is known; the revolutions per minute at which the engine is running can be ascertained by a mechanical speed counter; the area of the piston is known; all then we need to know is the mean effective pressure on the pistons, or what P stands for in the formula—

$$\frac{2SRAP}{33,000}.$$

The actual mean pressure exerted on the piston of an engine is found by means of an indicator. An indicator of the Crosby make is shown by Fig. 46. This instrument consists of a small barrel, or cylinder, in which a piston works against a spring. The piston which in a good indicator is nearly frictionless, actuates a pencil at the end of an arm or lever. When the indicator barrel is placed in communication with the engine cylinder, any pressure in the latter causes the indicator piston to rise, and the pencil marks a piece of paper or card. If the card were held stationary, a vertical line only would be recorded due to the rising and falling of the indicator piston, but by moving the card so that the portion under the pencil corresponds with the position of the engine piston, a continuous line or diagram is drawn showing the pressure exerted on the engine piston at all points of the stroke. The drum to which the card is attached is rotated by means of a cord attached to any reciprocating part of the engine which gives the required motion.

The Crosby indicator, owing to the lightness of its working parts and to the accuracy with which it is made, is largely used for indicating high-speed engines, for which purpose the older patterns of indicator having heavier parts and a longer stroke are unsuitable. The latest development of the Crosby indicator is shown by Fig. 46a, from which it will be seen that the spring against which the indicator piston works is placed outside the barrel. This is a great improvement, especially in cases where superheated steam is used, as the temperature of the steam does not affect the action of the spring. Moreover, when the spring is placed ouside the barrel, there is no possibility of its causing the piston to bear unduly on one side of the barrel.

Springs of different strengths are supplied with the indicator. Thus, one spring will travel upwards 1 inch with a pressure of 40 lbs. in the barrel, while another will travel 1 inch upwards with a pressure of 10 lbs. One is called a 40-lb. spring, the other a 10-lb. spring. The spring used is noted on the indicator card by the tester; a much stronger spring is used for indicating the H.P. cylinder than for the L.P. cylinder. To ascertain the pressure shown by a diagram, if a 40-lb. spring had been used, a boxwood scale graduated 40 lbs. to the inch would be used. If a 10-lb. spring had been used, a scale graduated 10 lbs. to the inch would be employed.

The indicator barrel is connected to the cylinder by means of a short pipe and cocks, arranged so that the barrel of the indicator may be shut off from the cylinder, and placed in communication with the atmosphere; when this is done, if motion is given to the card, a straight horizontal line is drawn, which is, of course, the atmospheric line.

Optical Indicators.—For speeds above 500 or 600 revs. per minute, an indicator with a piston, arm, and pencil, is unsuitable owing to the inertia of the moving parts. To indicate petrol or other engines running at very high speeds, optical indicators are used. In these the piston, lever, and pencil are replaced by a diaphragm and mirror. The movement of the diaphragm deflects the mirror, and a spot of light is thus thrown on a moving screen. With a fast-running engine, the spot of light traces what is apparently a continuous line, so that the shape of the diagram is seen while the engine is running. To obtain a record of the diagram, photography is resorted to.

Figs. 47 and 48 show two indicator diagrams taken respectively from the H.P. and L.P. cylinders of a 1,200 I.H.P. compound condensing Corliss engine supplied for driving a mill. At the time the diagrams were taken the engine was called upon to give between $\frac{2}{3}$ and $\frac{3}{4}$ of its full power only, but as the diagrams are reproduced from those of large size and are taken from a good engine, they will serve to show the use of such diagrams, and how the power of an engine can be worked out from them.

The explanation of the diagrams is briefly thus—The upper line represents the pressure of the steam in the cylinder on the outward stroke, the lower line the pressure on the return stroke. When steam is first admitted to the cylinder the indicator pencil rushes up to A; this initial pressure, in the diagram before us, is well maintained until the point of cut-off, when the steam begins to expand and the pressure falls. Point B shows approximately where the exhaust port begins to open; it is fairly wide open at C, and continues to remain open until D is reached when the exhaust port closes, the steam remaining in the cylinder begins to be compressed, the pressure rising to E. At this point the steam port again opens, and the pencil rushes up to A.

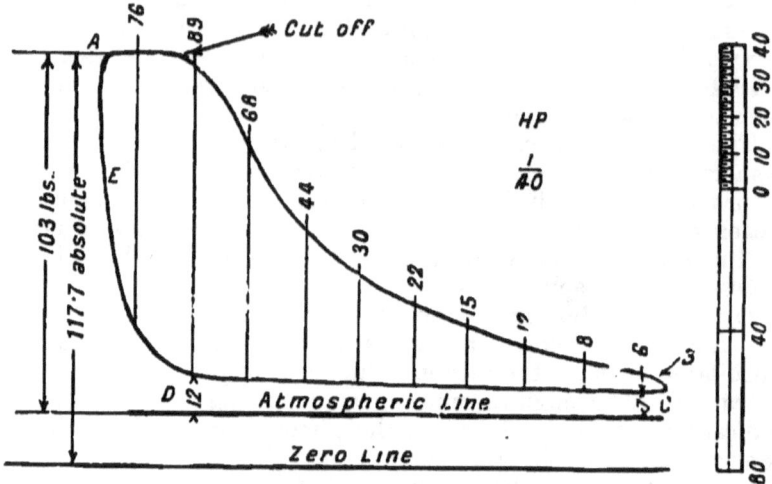

Fig. 47.—Indicator diagram.

The card taken from the other end of the cylinder was practically identical, but is not shown in order to avoid confusion.

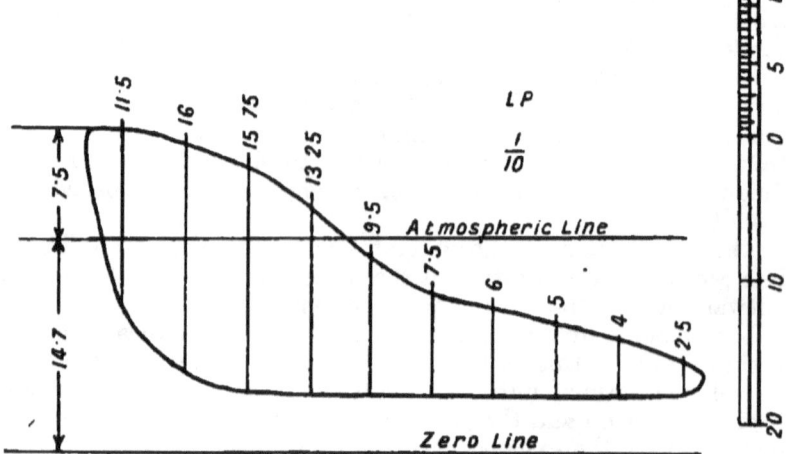

Fig. 48.—Indicator diagram.

The diagram shown by Fig. 47 was taken with a 40-lb spring, so that every inch of height represented 40 lbs. pressure on the piston. The diagram has been reduced in reproduction, but a scale is given at the side so that the reader can measure it for himself. To obtain the mean pressure shown by the diagram, an engine maker would use an instrument called a planimeter. By means of this instrument the area of any irregular space can be measured by merely running the pointer round the boundary line, when the area can be read off on the index; if, knowing the area, we divide it by the length of the diagram we get the mean height. However, as a planimeter is not possessed by every one, the next best way to find the mean pressure shown by the diagram is to divide it by 10 vertical lines, or ordinates, equally spaced. Then read off the height (or pressure in pounds) of each by the scale, add them all together, and divide by the number of lines. This will give the average or mean pressure.

The particulars of the engine from which the diagrams were taken are given below. The pressure shown by the boiler gauge was 112 lbs. above atmosphere or 126·7 absolute.

Stroke,	5	feet.
Revolutions,	60	per minute.
Diameter of H.P. piston,	30	inches.
Area of H.P. piston, after deducting area of 6-inch piston-rod and tail-rod,	678·5	,,
Diameter of L.P. piston,	56	,,
Area of L.P. piston after deducting area of piston-rod and tail-rod,	2,434·7	,,
Ratio of H.P. to L.P. cylinder,	1 to 3·59	

The mean pressures shown by the vertical lines are:—

H.P. diagram.	L.P. diagram.
76	11·5
89	16
68	15·75
44	13·25
30	9·5
22	7·5
15	6
12	5
8	4
6	2·5
370	91

If we divide the above totals by the number of lines—viz., 10—we get an average effective pressure of 37 lbs. per square inch on the H.P. piston, and 9·1 lbs. on the L.P. piston, or an equivalent pressure of 19·4* lbs. on the L.P. piston. We do not need to take any account of the back pressure against the pistons, as we have measured the height of the vertical lines not from the atmospheric line, or from the zero line, but from the line forming the bottom of the diagrams, which shows the pressure at which the exhaust steam leaves the H.P. cylinder to enter the receiver, and the pressure at which the exhaust steam leaves the L.P. cylinder to enter the condenser. Having obtained the mean pressure from the diagrams, we can now calculate the horse-power by the formula already given; the figures are—

$$\frac{2 \times 5 \times 60 \times 2{,}434\cdot 7 \times 19\cdot 4}{33{,}000} = 860 \text{ I.H.P.},$$

so that the engine was indicating 860 I.H.P. when the cards were taken. The H.P. diagram enables us to see what drop in pressure there was between the boiler and the cylinder. The boiler pressure was 112 lbs. above atmosphere, and the card or diagram shows that the highest pressure in the cylinder was 103 lbs. above atmosphere, or a drop of 9 lbs. The lower line of the diagram shows the pressure at which the steam leaves the H.P. cylinder; this pressure varies from 7 lbs. to 12 lbs. above atmosphere, say 9½ lbs. average pressure; the top line of the L.P. diagram shows that the highest pressure of the steam received in the L.P. cylinder was 7·5 lbs. The drop, which is not excessive, is caused by the steam having to pass through the exhaust port and receiver, and enter through the steam port of the L.P. cylinder.

The consumption of steam in the engine was only 14·51 lbs. per I.H.P. per hour; an extremely good result for a compound engine working at less than ¾ load. It was not possible to measure the actual load on the engine, so the mechanical efficiency or consumption per brake horse-power is not known as well as per indicated horse-power.

When it is not possible to measure the actual load, the approximate efficiency can be ascertained by running the engine light and taking diagrams; these will show the amount of steam required to turn the engine round. This rough efficiency test must be carried out before the driving ropes or belts are in position. Assuming the mean pressure required when the engine is fully loaded is 40 lbs., and the diagram shows a mean pressure of 4 lbs. when running the engine light, the engine efficiency is 90 per cent. This method is, however, not

* The L.P. piston is 3·59 times greater than the H.P. piston; a mean pressure of 37 lbs. on the H.P. piston is therefore equivalent to, say, 10·3 lbs. on the L.P. piston—10·3 + 9·1 = 19·4.

very reliable, as, in the first place, diagrams showing such a small mean pressure divided over the two cylinders are not very accurate, and, in the second place, the friction of the bearings and guides is greater at full than at light loads. It has been found that with high-speed engines constructed on interchangeable lines, and with the most accurate workmanship, the mean pressure required to run them light is considerably higher, and the consumption of steam greater, when first started than it is after the engine has had a few days' running; by which time the working parts will have acquired a fair working face.

When indicating a Willans' single-acting engine, an indicator card is always taken from the receiver beneath the H.P. cylinder (also beneath the intermediate cylinder, if there is one), and the mean pressure shown by such card is added to the mean pressure acting upon the upper side of the piston. The reason for this is as follows:—The lower line of the H.P. diagram shows the pressure of steam above the piston while the cylinder is exhausting to the receiver and the piston is on its up stroke; as soon as the piston reaches the top of its stroke the exhaust port closes. On the down stroke the diagram shows the pressure of steam driving the piston down, but it does not show what is going on beneath the piston; as a fact, the pressure beneath the piston—*i.e.*, in the receiver—begins to fall as soon as the L.P. cylinder commences to draw steam from it. This fall of pressure is shown by the receiver diagram, and is added to the pressure acting above the piston.

CHAPTER VII.

THE STEAM ENGINE.

(Part II.)

Simple Flywheel Calculations.—In addition to being able to work out the horse-power of an engine, it is essential that a young engineer should be able to make simple calculations required in connection with the flywheel, so that, for instance, he may be able to compare the value of two flywheels differing in size or weight, or to ascertain what effect the addition of a few inches to the rim will have upon the stored energy, and upon the tension in the rim due to centrifugal force.

The energy stored in a flywheel is expressed thus—

$$E = \frac{W \times V^2}{64 \cdot 33};$$

where W = weight of rim in lbs.
V = velocity in feet per second at the centre of the mass of the rim, or, as scientifically expressed, at the radius of gyration

Example.—What energy is stored in the flywheel of the engine shown by Fig. 35? We must first find the weight of the rim in pounds. There are two ways of doing this—one is to take from a table of areas the area of a circle corresponding with the outside diameter of the wheel, then take the area of a circle corresponding with the inside of the rim, and subtract one from the other. The result will be the superficial area of the rim. In the case in point the diameter of the wheel is 3 ; inches and the area 1,017 inches; the diameter inside the rim is 26 inches and the area 530 inches; subtracting one from the other we have 487 inches; the rim is 5 inches thick, so we multiply 487 inches by 5 inches, and the result is 2,43) cubic inches of iron. We read in Chapter i. that 1 cubic inch of cast iron weighs ·26 lb., so that if we multiply 2,435 by ·26 we have the weight of the rim in pounds—viz., 633 lbs. The other way is to find the circumference of a circle running through the centre of the rim,—97·4, and to multiply this by the sectional area of the rim—25 inches— the result is the same—viz., 2,435 cubic inches. Having got the weight, we must find out what V is. The engine runs at 300 revolutions per minute, or 5 revolutions per second; if, then, we take the circumference in feet of the wheel at the centre of the rim—viz., at the point indicated by a cross * in Fig. 35—and multiply it by the number of revolutions per second, it will give us the velocity at which the rotating mass is moving. The diameter of the wheel at the centre of the rim is 2 feet 7 inches, or 2·58 feet, and the circumference of a circle 2·58 diameter is 8·1 feet; if we now multiply the circumference by the number of revolutions per second, we get the answer 40·5 feet per second. By the formula we are required to square this; the square of 40·5 is 1,640. The calculation now is $\frac{633 \times 1,640}{64·33} = 16,137$. The stored energy is, therefore, 16,137 foot-lbs.

If the student will work out the energy of a flywheel having a rim of the same section—viz., 5 inches—but the wheel to be 6 inches larger in diameter, he will find that the stored energy has gone up greatly—viz., to 27,000 foot-lbs.

The diameter of a flywheel is governed by the peripheral speed at which it is safe to run. It is usually considered that a cast-iron flywheel should not run at a peripheral speed greater than 6,000 feet per minute; at this speed the bursting stress in the rim due to centrifugal force is only about 972 lbs. per square inch, and even if the engine is called upon to run at 10 per cent. above its normal speed, the bursting stress will only be about 1,176 lbs. per square inch. There is, of course, the possibility of the engine racing to be taken into account, but even should it race to the extent of 50 per cent., the bursting stress will be only about 2,186 lbs. per square inch, or a little under 1 ton. There is no doubt, therefore, that a peripheral speed of 6,000 feet per minute, or 100 feet per second, is a very safe one for a flywheel having a solid rim. If the flywheel is of such a size that the rim has to be in two or more pieces, then the safety, or

---

* To obtain this point accurately the following formula should be used:—

$$\sqrt{\frac{A^2 + B^2}{2}};$$

where A = outside diameter.
B = inside diameter.

This formula, which may perhaps look difficult to the beginner, is really very simple. It shows that we square the outside diameter of the flywheel, then square the inside diameter, add the two together, divide the product by two, and then find its square root from a table of squares. The calculation is as follows:—

$$\sqrt{\frac{36^2 + 26^2}{2}}, \text{ or } \sqrt{\frac{1,296 + 676}{2}}, \text{ or } \sqrt{986}.$$

A table of squares and square roots will show that the square root of 986 is 31·4, so that we should take the circumference of a circle 31·4 diameter instead of one of 31 inches diameter obtained by measuring to the actual centre of the rim.

otherwise, depends on the method of jointing, and a speed of 4,000 feet per minute is probably sufficiently high. With flywheels of very large size, say, 20 feet diameter and over, it is not usual to run at a peripheral speed much higher than 3,500 feet per minute.

Centrifugal force is a convenient expression, but what it really represents is covered by Newton's first law of motion. The law reads thus:—"*Every body continues in a state of rest or of uniform motion in a straight line* except so far as it is compelled by force to change that state." Now, every particle in the rim of a flywheel which has been set in motion tends to continue in a straight line, and unless force is employed to make the particle travel in a circle, the particle will continue to move in a straight line, which forms a tangent to the circle. The force which has to be employed to make the particle travel in a curved path is the equivalent of the centrifugal force exerted.

To find the centrifugal force of a revolving mass the following formula is used:—

$$C = \frac{W \times V^2}{32 \cdot 1 \times R};$$

where W = weight in pounds.
V = velocity in feet per second.
R = radius in feet.

The following will, however, be found a more convenient formula:—

$$C = W \times N^2 \times R \times 1 \cdot 226;$$

where W = weight in pounds.
N = number of revolutions per second.
R = radius in feet.

Example.—What is the centrifugal force of a piece of iron 1 inch square, rotating at 960 revs. per minute at a radius of 1 foot? The weight of 1 cubic inch of cast iron is ·26 lb., and 960 revs. per minute = 16 revs. per second; the square of 16 is 256. The calculation is, therefore, ·26 × 256 × 1 × 1·226. The answer is 81·6. The force required to make a piece of iron 1 inch square travel in a circle of 2 feet at the speed mentioned, instead of going off at a tangent, is 81·6 lbs.

Stress set up in the Rim of a Flywheel.—The formula given above enables one to find the centrifugal force of any portion of the rim of a flywheel, but to find the stress set up in the rim and tending to burst it as the result of centrifugal force of all its particles, it is necessary to find the centrifugal force of 1 cubic inch in the manner given above, then to multiply it by the diameter of the wheel, and divide the result by 2, or expressed as a formula—

$$B = \frac{C \times D}{2};$$

where B = bursting stress per square inch of rim.
C = centrifugal force of 1 cubic inch in pounds.
D = diameter of wheel in inches.

The centrifugal force is multiplied by the diameter, and not by the circumference of the wheel, for the same reason that the stress upon a cylinder wall under internal pressure is found by multiplying the diameter by the pressure. The reason is that the pressure on the circumference is not all equally effective, and the total effective pressure is equivalent to the pressure multiplied by the diameter; and, as in a cylinder the total pressure is borne by the two sides of a cylinder, so in a flywheel the total bursting pressure is borne by the two sides of the rim; hence, C multiplied by D is divided by 2.

The formula gives the stress on each square inch of the rim due to centrifugal force, and it is immaterial what the width may be. It is best to work out the stress in a ring equal to the outside diameter of the wheel and of 1 square inch section. The stress on the inner portion of the rim, assuming the latter is greater than 1 inch square, will, of course, be less than in the outer portion; but if a crack develops in the outer portion, the inner portion will give way also.

Example.—What is the stress in the rim of a small wheel, 2 feet diameter, running at 960 revs. per minute? From the example worked out above we have seen that the centrifugal force of 1 inch of the rim is 81·6 lbs.; we therefore multiply this by the diameter—viz., 24 inches—and divide by 2. The answer is 979 lbs. per square inch.

A very simple formula for finding the stress per square inch in the rim of a flywheel is given by Professor Unwin in his book on machine design. It is

$$\frac{3 \cdot 36 \, V^2}{32 \cdot 1};$$

where V = velocity in feet per second.

The figures 3·36 represent the weight of 1 foot of wrought iron of 1 inch area. To make the formula applicable to cast-iron flywheels, 3·12 should be used instead of 3·36.

An interesting feature about the centrifugal force acting on flywheels is this:—If we have a 20-feet wheel running at 95·5 revs., a 5-feet wheel running at 382 revs., and a 1-foot wheel running at 1,910 revs., the peripheral speed is the same in each case, but 1 cubic inch of cast iron placed on the rim of the large wheel exerts a centrifugal force of 8·1 lbs. only. One cubic inch of iron placed on the rim of the 5-feet wheel exerts a force of 32 lbs., while a similar piece of iron placed on the 1-foot wheel exerts a force of 162 lbs. The explanation is simple. It is this :—The path pursued by the piece of iron on the rim of the large wheel approaches more nearly to a straight line than is the case of the piece on the small wheel, and a smaller force is required to make it travel in the necessary curve. In the small wheel the departure from the straight line in a journey of 1 foot is much greater, hence greater force is required to hold it in.

The size of the wheel does not, however, affect the bursting stress per square inch of section of the rim, as, although in the 20-feet wheel the centrifugal force of 1 inch of metal is only about 8·1 lbs., there is the equivalent of 240 of such pieces to be reckoned with, while in the 1-foot wheel, although the force is 162 lbs., there are only 12 such pieces to be allowed for.

The above formulæ for finding the stress in the rim of a flywheel assume that the tension is all taken by the rim, the holding-in power of the arms being ignored. In flywheels of the disc pattern, as used on high-speed engines, the holding-in power of the disc is very considerable. If we take the case of a disc flywheel having a rim 6 inches wide by 6 inches deep, the disc 1 inch thick, the radius to the centre of the rim 1·5 feet, and the number of revolutions 660 per minute, we shall find that, although the speed is 6,200 feet per minute, the tension on the disc (due to the centrifugal force of a section of the rim 6 inches × 6 inches × 1 inch) is only 2,081 lbs. per square inch, so that even if the rim had a succession of saw cuts in it 6 inches deep, the disc would hold the pieces of the rim in place.

The author knew of a case where a 600 I.H.P. high-speed engine got out of hand while being run for the purpose of setting the governors; the engine finally attained such a terrific speed that it was completely wrecked, the only part remaining undamaged being the solid disc wheel.

Stored Energy in Flywheels.—It may be asked what amount of stored energy should a flywheel possess? The answer depends upon the type of engine to which the wheel is to be fitted, the work the engine has to do, and the regularity of turning that is required. For instance, a double-acting engine having three cranks placed at 120° apart will not require such a heavy flywheel as an engine with two cranks at 180° apart. An engine which has to drive a large circular saw and a dynamo will require a much heavier wheel than one driving several small lathes and a dynamo, for, in the case of the saw, the whole load comes off and on suddenly. Again, an engine which is required to drive a three-phase alternator in parallel with others will need a heavier flywheel than a similar engine to be used for driving the tools in an engineer's shop. To go deeply into the question of flywheel weights and fluctuations of speed, taking into consideration the varying pressures on the crank-pin due to steam pressure, inertia of moving parts, &c., is beyond the scope of this book, and unless the student is engaged with a firm of engine builders, he need hardly concern himself with such calculations. Even in engine builders' works the amount of stored energy of the flywheel is usually settled upon as the result of previous experience rather than as the result of calculation. For instance, it was found that with a certain type of single-acting three-crank engine, a flywheel having stored energy of 3,000 to 4,000 ft.-lbs. per I.H.P. was sufficient to meet the most onerous conditions, and a flywheel having this amount of energy was accordingly provided when the price obtained admitted of such a heavy flywheel.

In the huge flywheel, having a wrought-iron rim, fitted to the earlier 5,000 H.P. turbines at the Niagara Falls, the stored energy works out at about 22,000 lbs. per H.P., but the conditions necessitating a flywheel having such an enormous amount of stored energy are quite unusual.

Governing Steam Engines.—The speed of a steam engine is controlled by the governor and flywheel; the former controls the normal speed (sometimes called steady speed), while the latter prevents any sudden variations of speed, and gives the governor time to act. The governor may regulate the speed either by throttling the steam before it enters the valve chest, or by making the cut-off take place earlier or later during the stroke.

In the ordinary throttle governor the position of the balls determines the amount of opening of the throttle valve, and centrifugal force determines the position of the balls. The valve is set so that when the engine is running at full speed the passage of steam is unobstructed, or obstructed only so far as to give a little higher pressure on the boiler side. If the speed of the engine increases beyond the full speed for which it was designed, the balls open outwards and the throttle valve partially closes; if the speed of the engine falls, the balls close either by their own weight, or assisted by springs, and the throttle valve opens. When the engine is at rest the throttle valve is wide open.

In a governor which controls the speed of the engine by varying the expansion, as in a Corliss engine, the motion of the balls causes the catch plates to release the steam admission valve at an earlier or later portion of the stroke.

It should be clearly understood that a governor, however sensitive, cannot govern an engine to one absolutely uniform speed, as it is only by a variation of the speed that the centrifugal governor comes into play. A good governor should, however, be able to control the steady speed of the engine to within 3 per cent. under all changes of load, so that if the governor is set to give a speed of 100 revs. per minute under full load the engine should never run faster than 103 revs. even without any load. If the whole load is thrown off suddenly, the speed may rise momentarily beyond the 103 revs., but the governor will bring it back in a few seconds. The extent of the momentary variation of speed depends more upon the flywheel than upon the governor; therefore the flywheel should be sufficiently heavy to keep the momentary variation of speed down to 5 per cent. until the governor is able to produce its effect. When a complaint is made that an engine will not govern under sudden changes of load the fault may be looked for rather in the flywheel than in the governor.

A typical specification as to governing is the following:—
"The governor to be of an approved type capable of easy adjustment by hand while the engine is running. The governor to control the speed of the engine within 3 per cent. between full load and no load, and with a temporary variation of not more than 5 per cent. under any variation of load."

If an attempt is made to govern an engine much closer than is indicated by the above limits, or to make the governor "isochronous," there is a tendency for the governor to hunt. What is called hunting is this; when the governor is fitted with too sensitive springs and an increase of speed takes place, the throttle valve closes suddenly and thrott'es the steam to such an extent that the speed of the engine falls below the normal; the throttle valve then opens suddenly and allows rather too much steam to pass, when the speed rises above the normal; the governor again closes suddenly, and so the hunting action goes on. This hunting may give rise to very irregular running, and it is preferable to have a governor that will control the speed of the engine within reasonable limits, and be "stable," than to have one fitted with such light springs that it will hunt on the least provocation.

The balls of a governor should be fairly heavy, and controlled by strong springs if the spindle is horizontal, or by a weight if vertical. If the balls are light and the springs weak any friction of the governor mechanism may seriously interfere with the governing.

It is more difficult to govern within close limits a compound or triple-expansion engine than a simple engine unless the cut-off can be varied in each cylinder, as there is a certain amount of steam in the receivers which must be got rid of before the full effect of throttling or cutting off the admission in the H.P. cylinder is felt.

Some engines are fitted with an emergency governor. This governor does not come into action unless the main governor fails to hold the engine, and the speed assumes dangerous proportions. The governor of an engine should not be driven by a belt, as this may slip or break. In high-speed engines it is a common practice to fit the governor to one end of the crank shaft, and thus do away with any gearing; in cases where this is not practicable the governor should be driven by gearing.

Relative Economy of Governing by Throttling or by Varying the Expansion.—It is generally believed that it is more economical to govern an engine by altering the point of cut-off than to govern it by throttling down the steam at light loads. It is contended that if the boiler pressure is, let us say, 160 lbs., it is better to work with this pressure at all loads rather than throttle down the steam to 60 or 70 lbs. at very light loads. This is, in the main, correct, and with low admission pressure it is undoubtedly more economical to vary the expansion whatever the load may be than to further throttle down the steam; but when high initial pressures have to be dealt with the case is rather different, for if the expansion is carried out to too great an extent the range of temperature in the H.P. cylinder is very great, and the initial condensation is excessive.

If the load on the engine is so light that the steam is expanded until the mean pressure in the H.P. cylinder is about one-tenth of the admission pressure, then it is more economical to throttle down the steam and to expand less.

For any variations of load less than the above, the variable expansion type of governor is more economical than a governor of the throttling type.

In cases where an overload may come upon the engine, the variable expansion type of governor is far preferable to the throttling type. For instance, an engine is fitted with a throttle governor, and the full load is obtained with, say, 160 lbs. pressure and ·5 cut-off; if a sudden overload should come upon the engine it cannot respond, because the governor obviously cannot raise the steam pressure above that of the boiler, and the point of cut-off is fixed; whereas if the engine is fitted with variable expansion gear, as soon as the speed of the engine begins to fall the governor makes the point of cut-off later, and so increased power is obtained.

Proportions of, and Stresses in, Various Parts of a Steam Engine.—The following data, which are based upon present-day practice, may be useful:—

Ratio of Cylinders.—The ratios which the intermediate and L.P. cylinders bear to the H.P. cylinder are usually as follows :—

Compound Non-condensing Engines.		Compound Condensing Engines.		Triple Condensing Engines.		
H.P.	L.P.	H.P.	L.P.	H.P.	I.P.	L.P.
1	2·5 to 3·0	1	3·0 to 3·5	1	2·25 to 2·5	6·0 to 6·5

Area of Ports.—The area of the steam ports should be such that the flow of steam through them does not exceed 7,000 feet per minute. In the case of a locomotive running at 60 miles per hour, the speed of the steam through the ports is sometimes as high as 11,000 feet per minute, but at this speed the steam is seriously throttled. The area of the exhaust ports of Corliss engines is usually about 1½ times that of the steam ports. In the case of slide-valve engines the exhaust port is made from 1½ to 3 times the width of the steam ports. The exhaust port of the L.P. cylinder which communicates with the condenser should be made as large as the general design of the cylinder will admit of.

Having decided upon the speed at which the steam is permitted to pass through the ports, the area is found thus—

$$\text{Area of port in inches} = \frac{P \times S}{V};$$

where P = piston area in inches.
S = piston speed in feet per minute.
V = velocity in feet per minute at which steam is permitted to pass through port.

Thickness of Cylinder Walls.—The thickness of a cylinder wall is such that the stress upon it is about 1,500 lbs. per square inch, with a minimum thickness of ⅝ inch. The thickness (without reference to the minimum) can, therefore, be found as follows:—

$$\frac{D \times P}{3,000} = T;$$

where D = diameter of cylinder in inches.
P = highest admission pressure.
T = thickness of cylinder in inches.

The thickness of the H.P. and L.P. walls is usually the same.

Piston-rod.—This must be of sufficient size to withstand the alternate tension and compression due to the pressure on the piston (and to its inertia); it must also be sufficiently stiff to resist any tendency to bend or buckle under its load. The stress usually allowed is about 3,000 lbs. upon each square inch of the full diameter of the rod; the stress at the smallest part of the rod—viz., at the bottom of the thread cut for the nut used to hold the piston, the rod having previously been tapered down for the piston—may be as much as 5,500 or 6,000 lbs. per square inch. In locomotive practice the rod is usually enlarged where it fits into the piston, and before the taper begins; even then stresses up to 7,500 lbs. per square inch are often met with. The piston-rods of the H.P., I.P., and L.P. pistons are usually of the same diameter, and it will be found that this is usually from $\frac{1}{8}$ to $\frac{1}{10}$ the diameter of the L.P. piston. In large marine engines, or in engines where all the parts are light, the rod is often only $\frac{1}{10}$ the diameter of the L.P. piston, but in the case of Corliss engines and of compound locomotives the diameter is usually about $\frac{1}{8}$ that of the L.P. piston.

Connecting-rods.—The usual practice as regards length of connecting-rod is as follows:—

Marine engine,   3·5 to 5·0 times length (or radius) of crank.
Stationary
  land engines,   4·75 to 5·5   ,,   ,,   ,,
Locomotives,   5·5 to 7·0 *   ,,   ,,   ,,

The sectional area of the connecting-rod should be equal in its smallest part to that of the piston-rod, and the section should increase gradually towards the crank-pin end until it is $1\frac{1}{4}$ to $1\frac{1}{2}$ times the area of the piston-rod.

Eccentric-rods and Valve-rods.—The size of these depends largely upon the character of the valve gear, and no general rules can be given.

* In the case of a four-cylinder, ten-coupled locomotive recently constructed by the Austro-Hungarian Railway, the connecting-rod was 9·4 times the length of the crank.

Pressure on Journals, Crank-pins, and Crosshead-pins. — The pressure which may safely be allowed depends to a great extent upon the speed of the rubbing surfaces; thus a much greater pressure is usually allowed upon the crosshead-pin than upon the main bearings. The pressures usually allowed are as follows :—

Main bearings,	300 to 500 lbs.	per square inch.
Crank-pins,	600 to 900 ,,	,,
Crosshead-pins,	1,200 to 1,500*,,	,,

The above pressures are reckoned on the projected area of the bearing— *i.e.*, the diameter multiplied by its length, and not half the circumference multiplied by the length.

Diameter of Crank-shafts. — Crank-shafts of compound and triple-expansion engines of the marine type are usually made approximately one-fifth the diameter of the L.P. cylinder. Where there are two L.P. cylinders to one H.P. or intermediate cylinder, then the crank-shaft will be found to be approximately one-fifth the diameter of a cylinder having an area equivalent to the two L.P. cylinders. In the case of high-speed engines, where the reversal of stress in the shaft is very frequent, the crank-shaft is usually made about one-quarter the diameter of the L.P. cylinder. This rule does not apply to the crank-shafts of gas engines or to engines which have a heavy flywheel or rope pulley placed between the cranks, as is frequently the case with horizontal land engines.

Pressure upon Slide Bars.—The pressure upon the slide bars is found by multiplying the total pressure on the piston, by the length of crank, and dividing the result by the length of connecting-rod. There is usually no difficulty in providing slide bars and slipper having ample surface, and the pressures upon them are consequently light; they vary from 50 lbs. per square inch of surface in stationary engines to about 125 lbs. per square inch in the case of marine engines and locomotives. It may be stated here that friction theoretically is independent of the extent of rubbing surface, and depends only upon the pressure between the surfaces in contact. The coefficient of friction of dry steel to steel is about ·18, so that if we have a piece of steel weighing 100 lbs. pressing on another piece, it will require about 18 lbs. to move it about. If, however, we can keep a film of oil between the surfaces the friction is considerably reduced. The best way to ensure a film of oil remaining between the surfaces is to provide ample area, and consequently reduced pressure.

* Exceeded in locomotives.

Some experiments have been carried out at Cooper's Hill Engineering College * in order to ascertain the coefficient of friction between wrought iron and steel journals and bearings of different alloys, the journal and its bearing being immersed in a bath of oil. The following are approximately the results obtained with a steel journal running in a phosphor-bronze bearing :—

Pressure on Bearing per Square Inch.	Coefficient of Friction.	
	When Peripheral Speed of Journal = 400 Feet per Minute.	When Peripheral Speed of Journal = 100 Feet per Minute.
600 lbs.	·0032	·0029
500 ,,	·0035	·0029
400 ,,	·0040	·0030
300 ,,	·0045	·0033
200 ,,	·0060	·0040
100 ,,	·0095	·0054

It is probably impossible to ensure such a good film of oil remaining between sliding surfaces as between a shaft and bearing running in oil, as the shaft draws in oil in the same way that a pair of rolls draws in a sheet of metal. It is, however, advisable to provide large rubbing surfaces, so that the oil is not so easily squeezed out, and in this sense the statement that friction is independent of surface has to be qualified.

Piston Speeds.—The piston speeds usually found in actual practice are approximately as follows :—

Corliss engines,	. .	500 feet per minute.
Marine engines,	. .	500 to 1,200 feet per minute.
Locomotives, .	. .	1,120 feet per minute when running at 60 miles per hour.

Engine Packings.—In the early days of the steam engine, when steam pressures were low, it was customary to pack the glands of the piston- and valve-rods with hemp soaked in tallow, but with the advent of high pressures it was found that hemp charred, owing to the heat. When this occurred, the engine driver, in order to keep the gland tight, screwed down his gland tighter and tighter; the result was frequently a scored rod. Asbestos fibre packing was introduced about the year 1870, and

* See *Engineering*, vol. lxxxii., p. 595.

was found far superior to hemp for high-pressure steam, and it is very largely used at the present time. The method of packing a gland by squeezing any form of fibrous material up against the rod and walls of the gland is, however, open to objection. In the first place, if the driver screws down his gland nuts too hard, an excessive amount of friction on the rod is caused; it is possible to pull up an engine of about 50 H.P. by tightening the glands excessively. Secondly, the pressure on the rod is the same on both the steam and exhaust strokes. Thirdly, any slight lateral movement of the rod tends to cause the packing to leak.

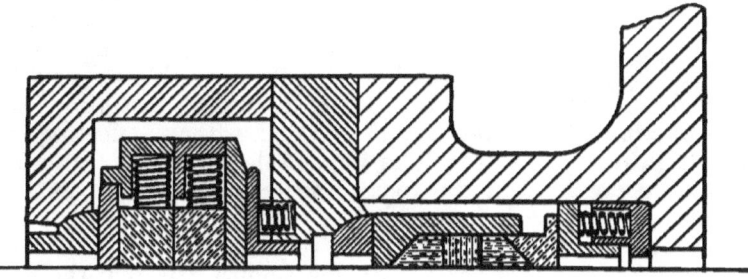

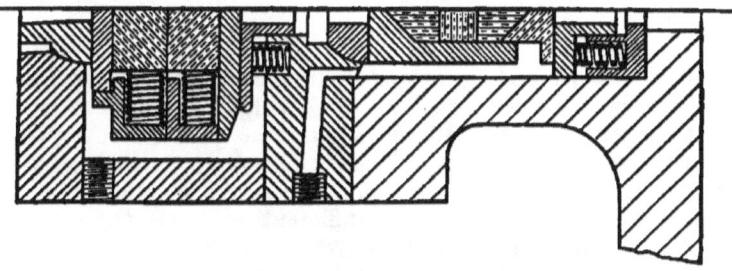

Fig. 49.—Metallic packing.

To overcome these defects various forms of metallic packing, consisting chiefly of soft white anti-friction metal rings or blocks, have been designed and patented. The best form of packing of which the author has any knowledge is that shown by Fig. 49. This packing was designed by a Scotch engineer, Mr. Monroe, patented in the United States, and is made and sold in this country under the slightly misleading name of the United States Metallic Packing. Fig. 49 shows the packing as supplied for very high pressures; for low-pressure cylinders the left-hand part of the packing only is required. The packing consists of eight white metal blocks, arranged so that they break joint; each block is pressed on to the rod by a light spring, so that the pressure on the rod is about $1\frac{1}{2}$ lbs. per square inch only; on the steam stroke, however, the pressure of the steam is added to that of the springs, and a tight gland is the result. On the exhaust stroke the pressure of the steam is removed, the pressure of the springs being sufficient to ensure the gland remaining tight under low pressure, or under atmospheric pressure when the engine is condensing. The white metal rings shown in the right-hand portion of the packing are for the purpose of reducing the pressure of the steam, when it is very high, before it reaches the packing blocks already referred to. As the reader will see from the illustration, the packing admits of lateral play of the rod without affecting the steam tightness; also, that it is impossible for the driver to put any undue pressure on the rod, as there are no gland nuts to tighten.

The form of packing described is very largely used in marine as well as in land work; the only objection to it is that the first cost is greater than that of a gland arranged to receive asbestos packing, and that occasionally the springs break. Against these objections it is claimed that, by reducing the friction on the rod, the mechanical efficiency of the engine is increased by 2 per cent. or more, and that the cost of replacing a spring is trifling.

Piston Rings.—In the *very* early days of the steam engine the piston was packed by means of rope or "junk," fitted into a recess, and held in position by a junk ring. The first to depart from this plan was Mr. Ramsbottom, who used narrow piston rings as shown by Fig. 52. These rings were turned slightly larger than the cylinder, were cut across, and sprung inwards; the spring in the material was sufficient to cause the rings to press tightly against the walls of the cylinder. Rings of this description are still largely used in locomotives and in fast-running petrol engines. There is nothing in them to go wrong, and they do not require a junk ring.

The rings are generally cut through at an angle, as shown by Fig. 50, so that there is not a direct path for the steam to blow through. When the rings are put into the piston the openings are spaced equally round the piston, but it is found that the rings have a tendency to work round into the position shown by Fig. 50. It is said that this tendency can be frustrated by cutting through the central spring at an angle inclined in the opposite direction to the angle of the others, as shown by Fig. 51. The objection to the Ramsbottom form of ring is that it

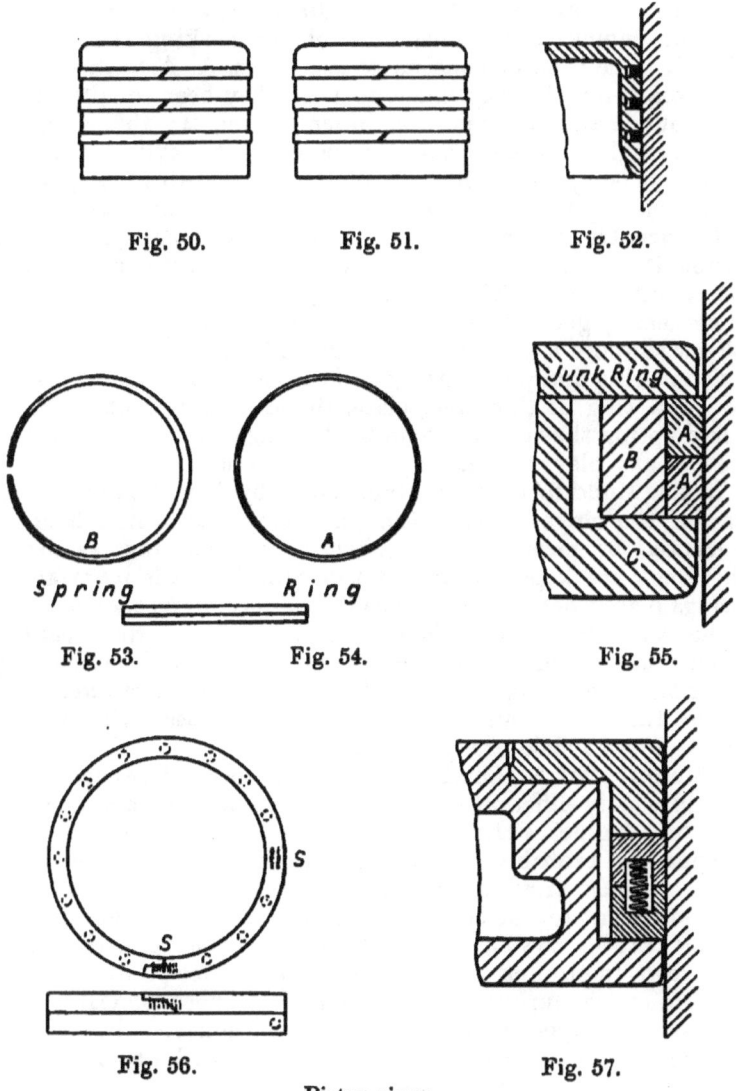

Fig. 50.   Fig. 51.   Fig. 52.

Fig. 53.   Fig. 54.   Fig. 55.

Fig. 56.   Fig. 57.

Piston rings.

cannot be replaced without drawing the piston, unless the latter is specially constructed, as shown by Fig. 38, and that, being originally made of slightly larger diameter than the cylinder, it does not press equally all round the latter when sprung in.

A simple form of piston packing, and one which has proved efficient in sizes up to 4½ inches diameter, is that used in the Willans engine; a sectional view of the packing is shown by Fig. 55. The packing consists of two rings, A, A, turned to the exact diameter of the cylinder, so that they bear equally all the way round, and one cast-iron inner spring, B; this spring is turned slightly larger than the bore of the cylinder, and is thicker at one portion than at another, as shown by Fig. 53; it is then cut, sprung inwards and placed inside the two thinner and concentric rings, A. The tendency of the spring B to resume its original form makes the rings A press evenly against the cylinder walls. The piston rings and spring require to be an extremely good fit between the junk ring and piston flange C, otherwise there is play, and the springs hammer themselves into the junk ring and piston flange. The effect was tried of making the junk ring of thin steel plate, the idea being that the steam would spring the steel plate on to the rings and spring, and thus prevent any play, but the experiment was not a success.

In the Mudd form of packing, shown by Figs. 56 and 57, any play of the rings between the junk ring and piston flange is prevented by means of a number of spiral springs as shown. The rings are kept up against the cylinder walls by means of springs placed between the ends of the former, as shown at S, S in the small scale plan of the ring, Fig. 6. This form of packing is used largely in marine work, and has proved satisfactory. There are a very large number of patent piston rings and springs besides those mentioned, but the examples chosen will serve, as well as any, to make the principles clear.

Efficiency of Steam Engines.—The term efficiency is often used in widely different senses. For instance, when a man says that his engine, although rather old-fashioned, is still very efficient, he probably means that the engine does not break down or cause trouble, and that he, the owner, is ignorant of, and indifferent to, the consumption of steam. Another man, in stating that his new engine is extremely efficient, probably refers to the consumption of steam per indicated horse-power; while a third, in endeavouring to sell a somewhat uneconomical engine, may lay stress upon its high mechanical efficiency. The mechanical efficiency of an engine, as already explained, depends solely upon the amount of internal friction.

The real efficiency, from the purchaser's point of view, is the consumption of steam at a given pressure (and temperature) per brake or effective horse-power.

The true thermal efficiency of an engine is the ratio the number of thermal units represented by the actual horse-power developed, bears to the number of thermal units put into the steam consumed by the engine. The efficiency of an engine reckoned in this way is so low (about ·157 in the case of a compound condensing engine, using 15 lbs. of steam per I.H.P. per hour, working with 150 lbs. boiler pressure) that it is seldom used in commerce, and is chiefly useful in comparing the performance of a steam engine with that of, say a gas engine.

Then, again, the thermal efficiency of a steam engine may be compared with that of a perfect steam engine working within the given limits of admission and exhaust temperatures. Such a standard has been recommended by the Institution of Civil Engineers (see vol. cxxxiv., p. 294 of these *Proceedings*).* The standard chosen is that laid down by Clausius and Rankine, and not that of the Carnot heat cycle.

Before concluding the chapter on steam engines, a few words must be said on the subject of the **Zeuner diagram**. This diagram enables the designer to see at what part of the stroke the cut-off takes place with a given amount of lap and lead; also the points at which the port opens to exhaust, and where compression begins.

Perhaps a little incident which actually occurred may show the beginner how the ability to construct such a diagram helped at least one young draughtsman a step forward. This young fellow, who wished to gain further experience, accepted a berth as draughtsman with a small firm of mechanical engineers on the coast. The firm had under construction a small marine engine of a size not previously made, and when it came to drawing out the slide-valve, the principal, who was an extremely good practical engineer, but who had not much theoretical knowledge, said to the draughtsman—"You had better make the lap three-quarters of an inch; I think that will be about right." The reply was—"Very well, sir, if you like I will set out a Zeuner diagram, so that we may see what the effect will be." The principal replied—"Oh, I think three-quarters of an inch will be near enough." However, after office hours, the draughtsman, for his own satisfaction, set out the Zeuner diagram, and the following morning showed it to his principal. The principal, whose guess as to the right amount of lap had been a good one,

---

* These volumes can be seen by anyone in London free of charge at the library of the Patent Office in Southampton Buildings, Chancery Lane. This library contains the past volumes of the *Engineer* and *Engineering*, and other technical papers; also a large collection of valuable text-books dealing with various subjects.

146   MECHANICAL ENGINEERING FOR BEGINNERS.

seemed very pleased to see set out so clearly the exact points where cut-off took place and where compression began, and at the end of the week the draughtsman was gratified to find that

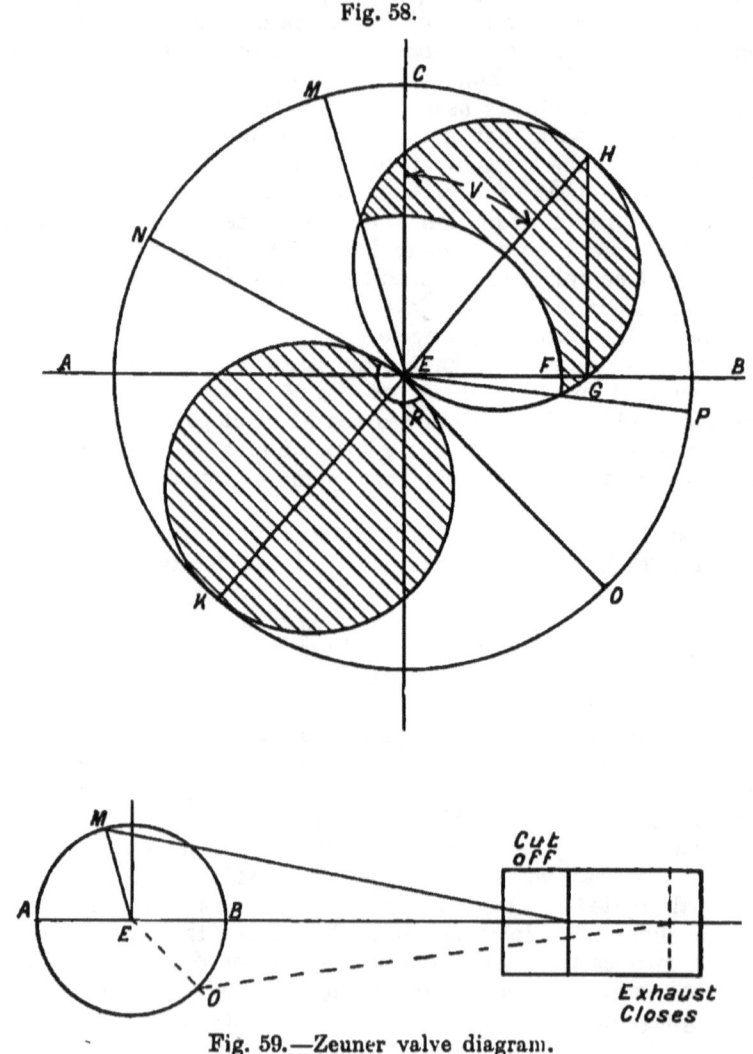

Fig. 58.

Fig. 59.—Zeuner valve diagram.

his salary had been increased. The principal, when thanked, merely said—"You are worth more than you are getting." Had the principal's guess turned out a bad one, it is just possible that he might not have been so pleased with the Zeuner diagram, but it would have been accurate just the same.

The diagram (Fig. 58) is constructed as follows:—Draw AB parallel to the line of the stroke. With a radius = $\frac{1}{2}$ the valve travel draw from the centre E the circle BCAK. Mark off EF = the lap, and FG = the lead. Draw the perpendicular GH. Join EH. Then V = angular advance. On HK draw the valve circles as shown. From E draw lap circles with radii EF = outside lap and ER = inside lap; then—

EB = position of crank at beginning of stroke.
EM =   ,,   ,,   at cut-off.
EN =   ,,   ,,   when exhaust opens.
EO =   ,,   ,,   ,,   ,,   closes.
EP =   ,,   ,,   ,,   steam port opens.

Having found the position of the crank at the points of admission, cut-off, compression, &c., one has merely to draw the connecting-rod and cylinder, as shown by Fig. 59, to obtain the corresponding position of the piston. In the illustration the connecting-rod has been assumed to be five times the length of the crank; EM is assumed to be the crank; therefore the connecting-rod is drawn five times the length of EM. One end of the cylinder will then be a connecting-rod's length from A and the other end a rod's length from B. By drawing the connecting-rod and piston, any position of the crank, found by Fig. 58, will give the corresponding position of the piston in the cylinder.

CHAPTER VIII.

POWER TRANSMISSION.

Belts, Ropes, and Gearing.—When an engine is not coupled directly to the machine it is required to drive, such as a dynamo, or, let us say, to a propeller shaft on board ship, and it is necessary to transmit the power to a machine some distance away, such transmission is usually effected by means of belts or ropes, pulleys, and steel shafting. For powers up to 100 H.P. flat leather belts are generally used; such belts can transmit up to 200 or 250 H.P., or more, but ropes are usually preferred when powers greater than 100 H.P. have to be dealt with.

The amount of power a belt is able to transmit depends upon the width of the belt, the speed at which it runs, upon the strength of the belt and its fastenings, and the extent to which the belt laps round and grips the pulley.

In a well-arranged horizontal belt drive the driving side is underneath, the slack side being uppermost, so that the sag of the slack side causes the belt to wrap itself more completely round the pulley than would be the case if the driving side were on the top and the slack side underneath. In the majority of cases of belt transmission the belts are placed at an angle. A very steep drive—*i.e.*, one in which the belt is nearly vertical—should be avoided if possible, as the more nearly the drive approaches the vertical the greater the tendency of the belt to slip. To avoid such slip the belt requires to be very tightly laced, and this causes undue friction on the bearings of the shaft, and consequent loss of power.

The power which can be transmitted by leather belts can be found approximately by the following formula:—*

$$\frac{TWV}{33,000} = H.P.;$$

---

* To be strictly accurate the formula should take into account the exact extent of the circumference of the pulley embraced by the belt, but in ordinary practice the width of a belt is never cut down to such a fine point that calculations going into these minute points need be made. A draughtsman who would spend a morning in making calculations of this nature in connection with the width of a pulley, would be of little use to his employer. The efficiency of a draughtsman is usually reckoned by the following formula:—

$$E = \frac{W \times A}{T};$$

where E = efficiency of draughtsman.
W = work turned out.
A = accuracy, sufficient for all practical purposes.
T = time occupied.

where T = working tension in pounds per inch of width.
W = width of belt in inches.
V = velocity in feet per minute.

The safe working tension of leather belts depends upon the thickness, and as the thickness of hide does not vary much, additional thickness is given by placing two or more hides together. A single belt is formed out of a thick single hide and is about $\frac{3}{16}$ inch thick; a double belt consists of two thicknesses of hide and may vary from $\frac{1}{4}$ to $\frac{3}{8}$ inch thick. For the purpose of the above formula the safe working tension of leather belts may be taken as follows:—

Single belt $\frac{3}{16}$ inch thick, T = 55 lbs.
Double belt $\frac{3}{8}$ inch thick, T = 80 lbs.

Let us work out an example.

Example.—What horse-power can be transmitted from a 3-foot pulley running at 200 revolutions per minute, the face of the pulley being 10 inches wide and the belt used, say, 9 inches wide? To find the velocity we multiply the circumference of the pulley, which is 9·42 feet, by the number of revolutions per minute, 9·42 × 200 = 1,884. We will assume that a single belt is used, the safe working tension of which we read above is 55 lbs. The calculation then is $\frac{55 \times 9 \times 1,884}{33,000}$ = 28·2 H.P. The answer is 28·2 horse-power.

If a double belt had been used, or if the velocity had been greater, the belt would have been capable of transmitting more horse-power.

If we wish to find the width of belt required to transmit a certain horse-power the formula transposed is

$$W = \frac{H.P. \times 33,000}{TV}$$

In cases where large powers have to be transmitted at low speeds, belts formed of leather links are sometimes used; these belts are more flexible than very thick solid belts; also as air can get away through the links they grip a wide pulley better than a solid belt, and are considered capable of transmitting about 30 per cent. more power. Centrifugal force, however, renders these belts unsuitable for high speeds, on account of their great weight.

Leather belts will run satisfactorily up to a speed of about 3,000 feet per minute; beyond this speed centrifugal force interferes somewhat with the gripping action. In cases where a belt will not transmit the required power, owing either to the result of centrifugal force, or to insufficient width, the required power can usually be obtained by using two belts, one over the other; this is sometimes called compounding. As the outer belt travels a little faster than the inner one and is usually narrower, the centrifugal forces do not act in the same way on the two belts and a better gripping action results. The power transmitted by two belts placed one over the other is greater than would be the case if the two belts were placed side by side on a very wide pulley, as apart from the centrifugal action one cannot be sure that the tension is the same in the two belts when so placed.

Leather belts when compounded have run satisfactorily up to speeds of 8,000 feet per minute.

Canvas belting is sometimes used for driving purposes; it is very flexible, but its life is not so long as that of a leather belt. The ends of leather belts are fastened together by laces. Many forms of metal fasteners are sold, but it is doubtful whether a joint made by them lasts as long as a properly laced joint.

Pulleys for driving belts are made of cast iron, wrought iron, and are sometimes built of wood. If the pulley is to be placed at the end of an engine or other shaft it is made solid and keyed on, but the majority of pulleys are now made in halves so that they may readily be put on the shaft while the latter is in position. Wrought-iron pulleys are slightly more expensive than cast iron, but are usually preferred on account of their lightness and strength. If the pulleys are to run at a high speed they must be properly balanced—$i.e.$, no part of the rim must be heavier than another unless exactly balanced by a corresponding weight on the opposite side. If this balance is neglected the pulley will run untruly, and may distort the shaft or loosen the bearing.

The faces of pulleys for leather belts, with the exception of fast and loose pulleys, are always "crowned"—$i.e.$, slightly convex; if they are not crowned the belt will run off. A leather belt always tends to mount the highest part of a pulley face; hence if the centre of the face is the highest part, the belt will run up to the centre of the pulley and remain there. If the convexity of a pulley is too great, the centre portion of the belt only grips the surface and the belt fails to transmit the power it is capable of. It has been found in such cases that by reducing the convexity of the pulley greatly increased power is transmitted. The rule frequently given in text-books that the convexity of a pulley should be from $\frac{1}{4}$ to $\frac{1}{2}$ inch for every foot width of pulley, if acted upon, would probably give rise to the trouble referred to.

In the opinion of the late Mr. Tullis, who probably had more experience in connection with belt driving than most men, a convexity of $\frac{1}{16}$ inch is sufficient for pulleys up to 6 inches wide, and a convexity of $\frac{3}{32}$ inch for very wide pulleys. If, however, the shaft is vertical and the pulley horizontal (an unusual arrangement except in the case of some machine tools), the convexity should be doubled.

By varying the size of the pulleys the relative speed of the driving and driven shafts can be varied through a wide range, but it is not good practice to have a greater difference than 4 to 1 between any two pulleys. That is to say, a 4-foot pulley should not drive one smaller than 1 foot diameter, or greater than 16 feet diameter; if this ratio is exceeded the extent of the circumference of the small pulley embraced by the belt or rope is small. When a large ratio must be given the pulleys should be placed as far apart as possible.

In cases where the two pulleys cannot be placed a good distance apart, and it is necessary for the belt to lap well round the small pulley, the belt is sometimes left rather slack, and a third pulley is used to press the slack portion of the belt towards the tight portion. This third pulley is called a Jockey pulley. It is not often employed, as it is noisy, and tends to wear the belt out quickly.

The speed at which a driven pulley of a given size will run, if the diameter and speed of the driving pulley are known, is found as follows:—

$$\frac{D \times S}{d} = s;$$

where D = diameter of the driving pulley.
S = speed of the driving pulley.
d = diameter of the driven pulley.
s = speed of the driven pulley.

Example.—Suppose we have a pulley, 3 feet in diameter, on a shaft running at 200 revs. per minute, which drives by belt a pulley 2 feet in diameter, at what speed will the latter run? By the above formula $3 \times 200 = 600 \div 2 = 300$. The answer is 300 revs. per minute.

If we know the diameter and speed of the driving pulley, and require to know what diameter the driven pulley must be in order to make the latter run at a given speed, the formula is merely transposed thus:—

$$\frac{D \times S}{s} = d.$$

Example.—Suppose the driving wheel is 3 feet in diameter, and runs at 200 revs. per minute, what diameter must the driven wheel be to give a speed of 120 revs. per minute? The calculation is $3 \times 200 \div 120 = 5$ feet.

POWER TRANSMISSION.

To ensure a long life for a belt, the pulleys should be of adequate size. In cases where the pulley must be of small diameter a thin belt should be used.

The following are suitable thicknesses of belts for small pulleys:—

Pulley 4 inch or less diameter; Belt $\frac{1}{8}$ inch thick.
" $4\frac{1}{4}$ to 8 inch " " $\frac{5}{32}$ "
" $8\frac{1}{4}$ " 12 " " " $\frac{3}{16}$ "
" $12\frac{1}{4}$ " 18 " " " $\frac{1}{4}$ "

In all cases where a machine is driven by belt from the main shafting, means must be provided for starting and stopping it without arresting the progress of the main shafting. This is effected by means of fast and loose pulleys; the pulley on the main shaft is a wide one, not crowned; the machine is provided with two pulleys side by side, one of which is free to revolve on its spindle. When the machine is at rest the belt drives the free pulley; when it is desired to start the machine the belt is moved by a fork and lever, called "the striking gear," on to the fixed pulley, and so the machine is driven.

In many cases, as in most lathes, it is usually more convenient to place the fast and loose pulleys on a shaft overhead. The countershaft, as it is called, has pulleys of different sizes corresponding with similar pulleys on the lathe, called cone pulleys; this arrangement permits of the speed of the lathe being varied: thus, when the belt is on the large pulley of the countershaft, it is on the small pulley of the lathe, and the latter runs fast. When the belt is on the small pulley of the countershaft, and on the large pulley of the lathe, the latter runs slowly. The fork of the striking gear prevents the belt from running off the uncrowned faces of the fast and loose pulleys.

Rope Driving.—The power transmitted by ropes may be found from the following figures, which err, if at all, on the safe side:—

One rope	$\frac{3}{4}$	1	$1\frac{1}{4}$	$1\frac{1}{2}$	$1\frac{3}{4}$	2	inches in diameter
Will transmit	2	2·5	4·5	6	8	10	horse-power if hemp ropes are used
	2·5	3	5	7	10	13	horse-power if cotton ropes are used
	for every 1,000 feet velocity per minute.						

Example.—Suppose we have a pulley, 38 inches in diameter, which has grooves for ten 1¼-inch ropes, the speed is 470 revs. per minute, what power will it transmit? We must first find the velocity: the circumference of a 38-inch pulley is 119·4 inches, or 9·95 feet; we multiply this by the number of revolutions, and the result is 4,676 feet velocity per minute. By the above rule one 1¼-inch cotton rope will transmit 5 H.P. for every 1,000 feet velocity; therefore, ten ropes will transmit 50 H.P. for every 1,000 feet velocity. The velocity we have found is 4·67 thousand feet per minute, therefore $4·67 \times 50 = 233·5$. The H.P. transmitted is 233·5.

A rope pulley should be not less than 30 times the diameter of the rope, if a cotton rope is used; or 40 times the diameter, if a hemp rope is used. Thus, if we have a pulley 30 inches in diameter we must not use cotton ropes larger than 1 inch in diameter, or hemp ropes larger than ¾ inches in diameter. A small rope bends more easily than a large one; therefore, as a general rule, it is better to have a good many small ropes than a few of large diameter.

Ropes may be run up to a speed of 7,000 feet per minute, but a speed of between 4,500 and 5,000 feet per minute is considered the best. Cotton ropes are more expensive than hemp ropes, but as they last much longer, they are really cheaper in the long run. Makers of ropes usually speak of them by the circumference; thus a maker's 3-inch rope would be one of about 1 inch diameter. Ropes should not touch the bottom of the V-shaped groove in the pulleys, but should wedge themselves against the sides. The best angle for the sides of the groove is about 40° when the diameter of the rope is over 1 inch, and an angle of about 30° when the rope is less than 1 inch in diameter.

When the diameter of a rope pulley is spoken of, the effective diameter—*i.e.*, the diameter of the pulley where the ropes grip the sides—is meant. The ends of ropes are spliced together; the rope makers send out men specially qualified for doing this work.

It has been found in practice that in cases where the drive is irregular, as with a gas engine, a steadier drive can be obtained by placing the slack side of the rope underneath, and the driving side uppermost, than by the opposite arrangement.

A question which frequently occurs in laying out a rope drive is:—What is the minimum distance at which the pulleys should be placed apart? A rule given by Mr. Kenyon (an authority on rope driving) is as follows:—Take the difference between the diameter of the largest and of the smallest pulley and add it to one and a-half times the diameter of the largest; the result gives the distance between centres of the pulleys.

Example.—We have a driving pulley 4 feet in diameter, and a driven pulley 1 foot in diameter; how near may they be placed together? The difference between the diameters of the two pulleys is 3 feet. We add this to one and a-half times the diameter of the large pulley—$3 + (1\frac{1}{2} \times 4)$ = 9 feet. The pulleys should therefore be placed not less than 9 feet apart. If the pulleys had each been 4 feet in diameter, the difference between their diameters would be *nil*, so that the minimum centres would be $1\frac{1}{2} \times 4 = 6$ feet.

When ropes are used for the transmission of power, ordinary fast and loose pulleys are useless, as the ropes cannot be passed from one to the other as is possible with a belt. To overcome this difficulty the pulley carrying the ropes can be made either fast or loose by means of a clutch. A plain pulley dog-clutch is one in which the pulley is free to revolve on the shaft, but adjoining it is a "dog" which cannot revolve on the shaft, but is free to slide along it. This dog has projections which, when it is pressed up against the pulley, engage with corresponding projections, and causes the pulley to revolve. When the dog is withdrawn the pulley remains stationary and the shaft revolves.

An expanding clutch is one in which one portion of the pulley carrying the rim is free to revolve, while the other portion, which is fixed to the shaft, is arranged so that its diameter can be increased or diminished. When the diameter of this portion of the pulley is increased, it grips the inside of the rim of the other portion, and so compels it to rotate.

Another clutch is similar in principle to the dog-clutch, but the dog is made of large diameter, and both it and the loose portion are provided with a large number of steel wire bristles similar to hair brushes. When the two portions of the clutch are brought together the bristles engage with one another, and the dog portion of the clutch compels the other portion to revolve. The object of this clutch is to avoid shock if the pulley requires to be coupled or uncoupled while running. The author's experience with this form of clutch is that it must be of very ample size for the work, otherwise the wire bristles do not last.

Loss of Power in Transmission.—The loss incurred in the transmission of power by means of belts, ropes, and shafting is considerable. It is usually considered that a belt drive absorbs about 5 per cent. of the power transmitted, and a rope drive about 7 per cent. In one case which came to the author's notice, a steam engine engaged in driving a portion of an engineer's machine shop, indicated 80 H.P. when all the tools were working. When none of the tools were at work, and the engine was merely driving the shafting and belts, the engine indicated 30 H.P. The power lost in engine friction was probably about 8 H.P., so that the remaining 22 H.P., or 27½ per cent. of the power developed, was spent in transmission.

This loss of power has often raised the question of electric transmission, but the high first cost of the dynamo and motors generally puts this method of transmission out of the question, even if the gain were shown to be considerable. In point of fact, the gain is not very great, as only 90 or 92 per cent. of the power put into the dynamo is given out as electric energy, and only 88 to 90 per cent. of the electrical energy is restored by the motor in the form of mechanical energy; there is consequently a loss of about 20 per cent., without taking into account the loss of power in connection with those tools which still require fast and loose pulleys and belts. If we assume that these absorb 7½ per cent. of the power, we are no better off through transmitting the power electrically than by doing so through shafting and pulleys. It would certainly be too expensive, even if it were desirable, to fit a small motor to every machine; the efficiency of small motors is nothing like so great as it is in those of large size. In cases, however, where power has to be transmitted to considerable distances, electric transmission is frequently advantageous.

Shafting.—The shafting which transmits the power and carries the pulleys is usually made of mild steel; the lengths are joined together by cast-iron couplings.

The power which may be transmitted by a steel shaft can be obtained from the following figures:—

1¼	1¾	2	2¼	2½	3	4	5	6	Dia. of shaft, in inches.
6·25	10	15	21	30	50	120	235	400	H.P. per 100 revs.

The power which a shaft will safely transmit varies directly as the speed; the powers given above are for a shaft running at 100 revs. per minute, so that, if the shaft runs at 200 revs., twice the power will be transmitted. If we wish to ascertain what power a 2½-inch shaft running at 225 revs. will transmit, it is only necessary to multiply 30 by 225 and divide by 100; the answer is 67·5 H.P.

The above figures will be found to correspond very nearly with the best present-day practice, but in any extreme case common-sense must be used. For instance, if we wish to transmit 6¼ H.P. through a shaft several hundred feet long running at 100 revs., we should probably decide to use a 1¾-inch or even a 2-inch shaft (rather than one 1½ inch diameter) for the sake of stiffness, and to prevent the shaft whipping. Whereas, if the power had to be transmitted through a shaft a few feet long, a 1½-inch shaft would be sufficient. The strength of a shaft varies as the cube of its diameter, so that if the reader wishes to ascertain how much power can be transmitted through any shaft, the diameter of which is not given above, it is easy for him to do so. Suppose we wish to know how much power can be transmitted through a shaft 10 inches in diameter, running at 100 revs. per minute. We see that a 6-inch shaft will transmit 400 H.P. at this speed; the cube of 6 inches is 216, and the cube of 10 inches is 1,000; we therefore have a simple proportion sum—thus, 216 : 400 :: 1,000 : x. The answer is 1,850; a 10-inch shaft will therefore transmit 1,850 H.P. at 100 revs., 3,700 H.P. at 200 revs., 4,625 H.P. at 250 revs., and so on. The rule gives results a little on the safe side, especially in the case of large shafts; for instance, the engineers of the first turbine installation at Niagara provided a 10-inch shaft next the turbine for transmitting 5,000 H.P. at 250 revs., whereas by our rule we should only have allowed 4,625 H.P., or have made the shaft $10\frac{1}{4}$ inches diameter. However, the shaft journals some distance away from the turbines were made 11 inches diameter, probably for the sake of stiffness.

The figures given apply to shafts which are subject to torsion in one direction only, and the reader is warned that he must not apply the rule to propeller shafts or to the crank-shafts of steam and gas engines. These shafts are subject to shock and stresses which are not easily calculated; the only safe guide as to the right proportions of a crank-shaft is actual experience. A well-known firm of engineers, making a very successful gas engine of large powers, has found it necessary, on account of breakages, continually to increase the diameter of its crank-shafts beyond the sizes which calculation would appear to render necessary, until the crank-shafts are now half the diameter of the piston! A somewhat similar experience befell a firm of high-speed engine makers, who many years ago suffered from broken crank-shafts. The directors of this firm were told by the work's manager that the shafts were too weak, and that they ought to be one-quarter the diameter of the L.P. piston. The manager was told that such a rule of thumb was perfectly ridiculous, and that the stresses could be calculated without any difficulty. After many breakages, further calculations were made, enormous factors of safety being apparently allowed, the shafts were strengthened, but still they broke. The shafts were again still further strengthened, and finally the breakages ceased; but, curiously enough, the dimensions of nearly all the steam-engine shafts now made by this firm approximate very nearly indeed to the works' manager's despised rule.

On the face of it, a rule which does not take into account the mean pressures, inertia of working parts, length of stroke, &c., does look ridiculous, but, upon going more closely into the matter, it will be found that one factor is probably counterbalanced by another; for instance, the L.P. piston of a non-condensing engine, having a late cut-off, may have a much greater mean pressure upon it than the L.P. piston of a condensing engine having a fairly early cut-off, but the former will have a better cushion to absorb the inertia of the working parts than the latter. Again, at first sight, one would say that a slow-speed engine, the L.P. cylinder of which is 20 inches in diameter by 30 inches stroke, would require a larger crank-shaft than an engine having the cylinder of the same diameter, but with a 10-inch stroke, as not only is the crank of the former engine three times longer than that of the short-stroke engine, and the torque on the shaft correspondingly greater, but the weight of the parts of the long-stroke engine will also be greater, owing to the greater length of connecting-rod. Against these considerations must be set the fact that the rapid alternations of stress in the crank-shaft of an engine, running at the high speed which would be expected from an engine having a 20-inch by 10-inch L.P. cylinder, are much more punishing to the shaft than any stress set up in the shaft of the slow-speed engine. As a fact, in actual practice, the crank-shafts of long-stroke, slow-speed steam engines are usually made about one-fifth the diameter of the L.P. cylinder, while the crank-shafts of short-stroke, high-speed engines require to be made about one-fourth the diameter of the L.P. cylinder, if they are to have a reasonable life.

Bearings.—The bearings which carry the shafting are called plummer blocks; they contain a top and bottom brass which can be renewed when much worn. The plummer blocks are usually carried on cast-iron brackets; when suspended from a joist or beam the brackets which carry the plummer blocks are called hangers. It is important that all plummer blocks should be accurately aligned, and that the brasses should be properly lubricated in order to minimise friction as much as possible. Hangers are sometimes made with adjustable devices, by means of which the bearing may be raised, lowered, or made to swivel.

Gearing.—Before the advent of rope driving, it was usual to transmit large powers by means of gearing. For instance, a large mill engine would transmit by gearing its power to a vertical shaft running the whole height of the mill; on each floor the power would be transmitted to horizontal shafts by means of bevel wheels; the horizontal shafts would then drive the machinery by belts.

The loss of power in transmission by gearing is less than in transmission by ropes, but gearing is noisy and cumbrous, and a broken toothed wheel may involve a serious stoppage of the mill. Transmission of large powers by gearing is now seldom resorted to, except in cases where large powers have to be transmitted at low speeds, as in steel rolling mills.

Toothed gearing is, however, very useful in transforming small powers at high speeds to great powers at reduced speeds, as in the case of cranes. Thus, if a small toothed wheel or pinion, 2 inches in diameter, drives a toothed wheel, 20 inches in diameter, the latter will be able to raise by a drum a weight ten times greater than would be possible if the weight were being lifted by a drum fixed directly to the 2-inch wheel. The weight will, however, be lifted ten times more slowly, so that what is gained in power is lost in speed.

It is essential that the teeth of gear wheels should be machine cut and accurately formed, if the gear is to work quietly and to waste little power.

Toothed wheels are called spur wheels when their shafts are parallel and they drive in the same plane. When a small wheel drives a large one the small wheel is called a pinion.

When one wheel drives another at right angles to it, or at an angle slightly greater or less than a right angle, the wheels are called bevel wheels.

When the sides of a wheel are carried up so as to support the ends of the teeth, the wheel is said to have shrouded teeth. But one wheel only out of a pair can have shrouded teeth if carried up to the top of the teeth. Both wheels can have shrouded teeth if the shrouding ends just below the pitch line. A wheel with shrouded teeth is stronger than one with plain teeth, but the teeth cannot be machine cut.

Helical wheels are those in which the teeth, instead of running straight across the face, are placed so that every tooth forms two sides of a triangle, the apex of the triangle being at the centre of the face. The effect of this is to give more surface, so that a wheel of a given width with helical teeth will transmit more power than one with plain straight teeth, and the action is smoother. Teeth of this form cannot be machine cut unless the two halves of the teeth are separated by a small space.

Worm Gearing.—Fig. 60 shows a worm and wheel; the worm has a single spiral, or thread, with a 1-inch pitch—*i.e.*, 1 revolution of the spiral will move forward the teeth engaged with it 1 inch. The wheel has twenty teeth of the same pitch, so that it will take 20 revolutions of the worm to make the

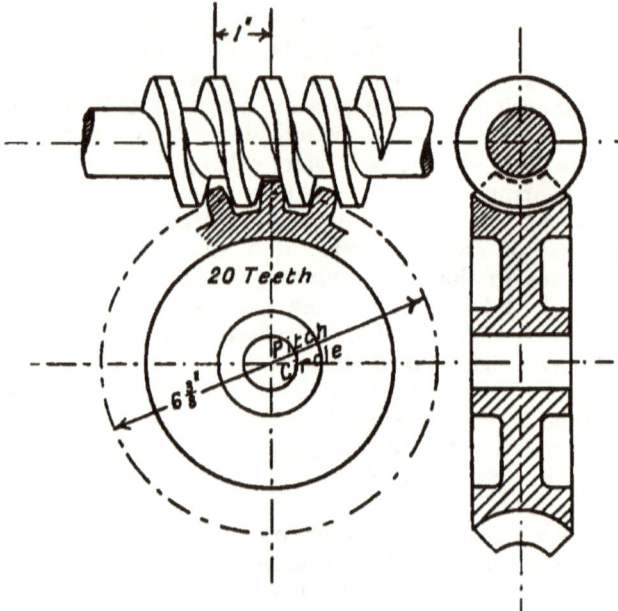

Fig. 60.—Worm wheel.

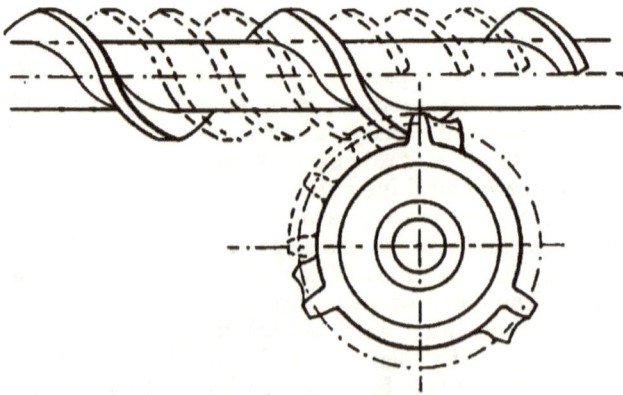

Fig. 61.—Worm wheel.

wheel revolve once. With a single spiral it is not practicable to effect a reduction of speed greater than 15 to 1, and 20 or 25 to

1 is a better proportion. Such a reduction is too great for most purposes.

Fig. 61 shows what the effect would be of a 3 to 1 reduction with an ordinary worm. The pitch of the worm would have to be very great, so that, for every revolution, it would turn the wheel through one-third of a revolution. The result would be that the worm would drive one of the teeth until it was out of mesh, but the next tooth would not then be far enough round to engage. The difficulty can, however, be got over by putting more spirals on the worm, as shown by dotted lines, and making a corresponding addition to the number of teeth on the wheel. The pitch of the spirals would remain the same, and the reduction of speed would also remain the same.

With the ordinary worm and wheel, as shown by Fig. 60, the wheel will not drive the worm ; but, with a worm having a very coarse pitch, the wheel will drive the worm and *vice versâ*. Worms with a coarse pitch and several spirals are used in motor cars where the worm drive is employed, otherwise the car could not move without the engine being turned by hand or being run by petrol. With the worm drive it is important to make the effective contact surfaces as large as possible, so as to get a large area of lubrication.

Skew wheels are constructed on the same principle—*i.e.*, a worm having many spirals, or, rather, parts of spirals—as the skew wheel is not sufficiently wide for a single thread to run right round it, as in the case of a worm; this, of course, is immaterial so long as the pitch is right.

Gear wheels, where great strength is required, are made of steel, and in high-class work the teeth are cut out of the solid. In large slow-running wheels the teeth are sometimes cast of the desired shape and trimmed up by hand. Toothed wheels for machine tools are usually made of cast iron and carefully machined. Such wheels work more smoothly than steel wheels.

Within recent years raw-hide pinions have been introduced to get over the noise and jar caused by a pair of wheels running at high speeds; they appear to answer admirably. Before these were introduced, the author was present at an attempt made to drive by gearing a dynamo which absorbed about 60 horse-power and required to run at 900 revolutions per minute, this speed being, of course, too high to admit of coupling the dynamo shaft directly to the engine shaft. The engine ran at 380 revolutions per minute and drove the dynamo through a pair of machine-cut steel wheels, but the noise made was so appalling that this method of drive had to be abandoned.

A good method of transmitting moderate powers at fairly high speeds in cases where belts are not admissible, and gearing is too noisy, is by the Hans Reynolds chain, shown by Fig. 62. With this chain a perfectly vertical drive is permissible. At a large engineering works in the Midlands, each main line of shafting running down the works is driven by this chain from a motor placed directly underneath the shafting. The motors run at a speed considerably higher than that of the shafting.

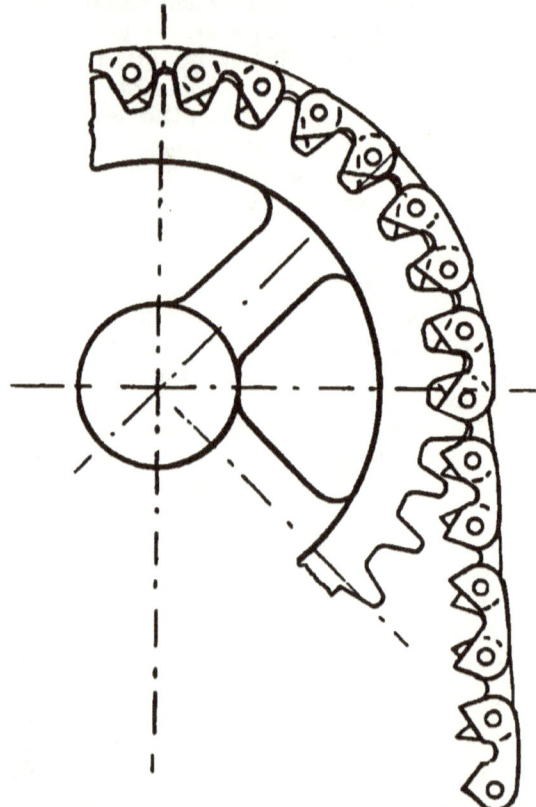

Fig. 62.—Hans Reynolds' silent chain.

With regard to the horse-power transmitted by gearing, it is not possible to give a simple rule, as in the case of transmission by belts and ropes. A formula often used, and which was originally published by Messrs. Musgrave, is as follows:—

$$\text{H.P.} = P^2 \times B \times V \div 1{,}000 \text{ for cast iron,}$$
$$\text{H.P.} = P^2 \times B \times V \div 625 \quad \text{,,} \quad \text{steel;}$$

where P = circumferential pitch of teeth in inches.
B = breadth of wheel in inches.
V = velocity of pitch line in feet per minute.

This formula, however suitable for toothed wheels of fairly large size, is not suitable for small wheels—*i.e.*, in cases where the pitch is 1 inch or less. Let us see how it applies to a small toothed wheel used in the gear-box of a 15-H.P. motor car. The wheel we will take has a $6\frac{3}{8}$-inch-diameter pitch circle; it has twenty teeth of 1-inch pitch, its breadth is $1\frac{1}{4}$ inches, and the velocity of the pitch line is 1,500 feet.

$$\frac{1^2 \times 1\cdot 25 \times 1{,}500}{625} = 3 \text{ H.P.,}$$

so that by the formula the wheel would only transmit 3 H.P.

In actual practice the wheel transmits 15 H.P. or more. An empirical formula, given by Mr. Box, for the strength of teeth is as follows:—

$$S = P \times W \times 350,$$

where S = safe load on one tooth in lbs.
P = pitch of wheel in inches.
W = width of tooth „

If we apply this to the wheel under consideration, we get $1 \times 1\cdot 25 \times 350 = 437$ lbs. safe load on one tooth. Let us see what power this will give us. $\frac{437 \text{ lbs.} \times 1{,}500 \text{ feet}}{33{,}000} = 19\cdot 8$ H.P., so that, by this rule, the wheel which is used on a 15-H.P. motor car will safely transmit 19·8 H.P. This formula is unsuitable for wheels in which the pitch of the teeth is much greater than 1 inch.

A beginner, for whom this book is intended, is hardly likely to be called upon to design gear wheels for some time, and, as a fact, there are in most drawing offices where such wheels need to be designed some available data as to sizes, strengths, &c.; such data, if intelligently used, are of far more value than any empirical formulæ. If in doubt as to the strength of a gear wheel, it is a good plan to assume that one tooth must be sufficiently strong to transmit the whole of the power, and to see what pressure in pounds will come upon it. This pressure will, of course, be found by the following formula:—

$$\frac{\text{H.P.} \times 33{,}000}{V};$$

where V is the velocity in feet per minute of pitch circle.

Knowing the section of the tooth, it is not difficult to form an opinion as to whether it is strong enough to bear the load.

CHAPTER IX.

CONDENSING PLANT.

CONDENSERS, although differing in type, are all designed with the same object—viz., to extract from the steam the heat remaining in it after the former has done its work, so that it condenses and a vacuum is formed, thus relieving the engine or turbine from the necessity of discharging the steam against atmospheric pressure. Means must be provided for getting rid of the condensed steam without permitting air to get into the condenser, and so impair the vacuum.

The condensers most frequently used are either jet or surface condensers. In an ordinary jet condenser the cooling water is admitted in the form of a jet or spray, as shown by Fig. 63. The exhaust steam from the engine, coming into contact with this spray of cool water, immediately condenses, and a vacuum is created. When the vacuum is once formed, the condenser will continue to draw in the injection water (owing, of course, to the pressure of the atmosphere on the surface of the water outside), so that, unless the water has to be lifted more than 10 or 12 feet, a pump to supply the condenser with injection water is not required.

With a jet condenser, in which the water mixes so intimately with the steam, a smaller quantity of water is required to condense the steam than is necessary in a surface condenser where the water has to effect its cooling action through tubes; but, on the other hand, if it is desired to use the condensed steam over again for feeding the boiler, it is necessary for the cooling water, as well as the feed water, to be free from impurities which may be injurious to the boiler, for, as we have said, the injection water mixes with the exhaust steam.

The amount of injection or cooling water required for a jet condenser is about 25 or 30 times the weight of the steam to be condensed. The size of a jet condenser is not of great importance; it is usually made about three-quarters the capacity of the L.P. cylinder. The shape, too, is of but small importance.

With a jet condenser fixed at, or below, the level of the engine, a pump is required to remove the condensed steam, cooling water, and a certain amount of vapour from the condenser; such a pump is called an air pump, as it prevents the air from getting into the condenser, also to distinguish it from a circulating pump.

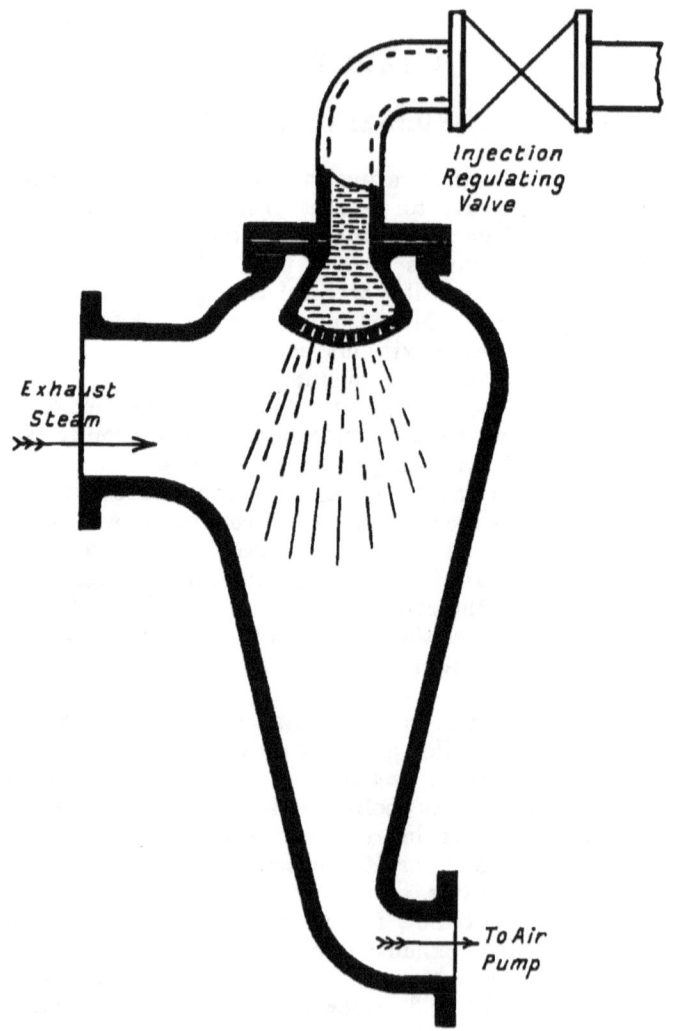

Fig 65.—Jet condenser.

An air pump is shown by Fig. 64. The pump is shown with head valves, bucket valves, and foot valves, but in practice the foot valves are now often dispensed with. The Edwards air pump, which has neither bucket valves nor foot valves, will be described later.

Surface Condensing Plant.—In a surface condenser the cooling water is kept separate from the steam which it has to condense. The water is passed through a large number of tubes, usually about $\frac{3}{4}$ inch outside diameter, made of Muntz metal or of an alloy composed of 70 per cent. copper and 30 per cent. zinc. A surface condenser is shown by Fig. 65. A condenser of this

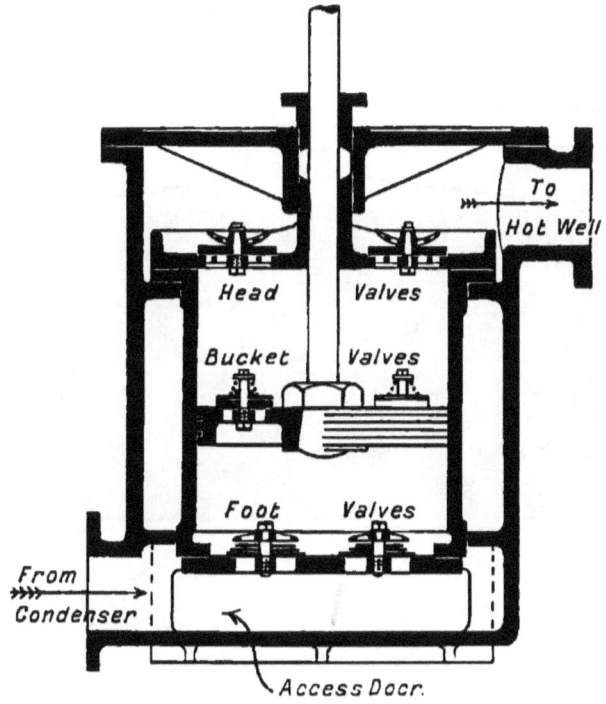

Fig. 64.—Air pump.

type is always used on board ship, where it is necessary to use the condensed steam over and over again for feeding the boilers, and where it is inadmissible to mix sea water with the feed water. A surface condenser is also used on land in cases where the cooling water is not suitable for use in the boilers.

With a surface condenser the air pump may be considerably smaller than with a jet condenser, as it has to deal with the condensed steam and vapour only; a separate pump, called a circulating pump, is used for circulating the water through the tubes. The circulating pump may be similar to the air pump shown by Fig. 64; but pumps of the centrifugal type, in which there are no valves, are frequently used. In a centrifugal pump (illustrated in a later chapter) there is an impeller, which is rotated at a rapid rate, and imparts sufficient motion to the water to make it travel through the tubes of the condenser.

Extent of Cooling Surface.—In a surface condenser there should be 1 square foot of cooling surface for every 9 to 10 lbs. of steam to be condensed per hour, assuming the temperature of the cooling water to be about 60° to 70° F. In the case of condensing plant for a steam turbine where it is desirable to get

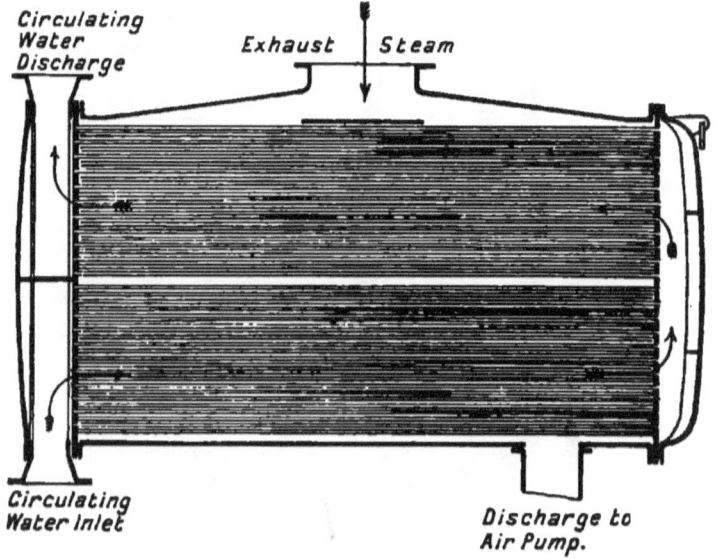

Fig. 65.—Surface condenser.

an extremely good vacuum, a square foot of cooling surface is frequently provided for every 6 to 8 lbs. of steam to be condensed.

The tubes of a surface condenser should be of a length not exceeding 12 feet, unless supported in the centre. Each tube is free to slide in its hole in the end plate, the joint being made by a small brass ferrule and cotton rope packing, as shown by Fig. 66. If the tubes are screwed or expanded into the end plates they are not able to expand or contract, and the condenser does not remain air tight.

It has recently been found that by dividing a surface condenser into sections horizontally and draining away the water from each section, a smaller cooling surface is equally effective. It is believed that a thick film of water hangs round the tubes, and prevents the conduction of heat through them. If the water is drained away from the upper rows of tubes as it is formed, the lower tubes remain fairly dry, and are much more effective.

Amount of Cooling Water Required.—The amount of cooling water required in connection with a surface condenser is variously stated in engineering pocket-books and text-books to be from thirty to seventy times the amount of feed water.

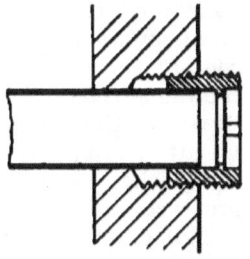

Fig. 66.—Tube end and ferrule, half size.

The amount of cooling water required depends on its temperature, but the following table, showing the actual vacua obtained with different quantities of feed water at a temperature of 65° F. may be useful. The table has been prepared from curves given by Mr. Allen in his paper upon condensing plants read before the Institution of Civil Engineers in 1905.

The curves themselves were plotted from a very large number of experiments carried out by Mr. Allen:—

Cooling Surface = 1 Square Foot for 5 Lbs. of Steam Condensed.		Cooling Surface = 1 Square Foot for 10 Lbs. of Steam Condensed.	
Amount of Cooling Water.	Vacuum.	Amount of Cooling Water.	Vacuum.
40 times feed.	27·2″	40 times feed.	26·75″
50 ,,	27·7″	50 ,,	27·25″
60 ,,	28·0″	60 ,,	27·75″
70 ,,	28·25″	70 ,,	28·0″

From the above figures it will be seen that for a vacuum of 26 inches or 27 inches, forty times the amount of the feed is sufficient, but if very high vacua are desired the amount of cooling water must be increased, or water of a lower temperature used.

The air pump in the experiments in question had a capacity of ·75 cubic foot per pound of steam condensed. Temperature of cooling water 65° F., Barometer 29·9.

Capacity of Air Pump.—The capacity of the air pump—*i.e.*, the volume swept by the bucket multiplied by the number of

170 MECHANICAL ENGINEERING FOR BEGINNERS.

strokes per minute—should be from ·75 to 1 cubic foot per pound of steam to be condensed. With some surface condensing plants a pump capacity of 1·5 cubic feet per pound of steam to be condensed has been allowed; it is, however, useless to provide an air pump of very large capacity, unless the temperature of the cooling water is fairly low, and the condenser has adequate cooling surface. For example, assuming the condenser is unable to reduce the temperature of the exhaust steam below 105° F., no air pump, however large, can give a better vacuum than 27·7 inches (see the subjoined table), for at this temperature and pressure water turns into steam. If an attempt is made to get a higher vacuum than this by means of the air pump, not only will the exhaust steam not condense, but any water lying in the condenser will vaporise, and, as the pump will be quite unable to cope with such an enormous volume of steam, the vacuum will fall to a point at which steam will condense at a temperature of 105° F.

TABLE XVI.

Degrees Fahr.	Vacuum in Inches when Atmospheric Pressure = 14·7 Lbs.	Absolute Pressure in Inches of Mercury.	Absolute Pressure.	Total Heat in 1 Lb. of Steam from 32° F.	Volume of 1 Lb. of Steam.
			Lbs. per sq. in.	B.T.U.	Cubic feet.
32	29·7	·181	·089	1091·2	3,226
50	29·6	·362	·178	1095·6	1,695
60	29·4	·517	·254	1099·7	1,220
70	29·2	·733	·360	1102·8	877
80	28·9	1·024	·503	1105·8	641
90	28·5	1·410	·693	1108·9	549
100	28·0	1·917	·942	1111·9	353
105	27·7	2·229	1·095	1113·4	307

From this table it will be seen that, in order to obtain a vacuum of 28 inches, the temperature of the steam must be reduced below 100° F. To obtain a vacuum of 29·6 the steam must be cooled below 50° F.

If the atmospheric pressure is higher than 14·7 lbs. (or 29·95 inches of mercury) then a slightly better vacuum can be obtained with the above-mentioned temperatures. The vacuum it is possible to obtain theoretically with given temperatures, and with the barometer standing higher or lower than 29·95 inches can be ascertained by deducting the pressure given in inches of mercury (column 3) corresponding with the temperature of the steam, from the height of the barometer in inches of mercury.

Some engineers employ two air pumps with a surface condenser, one a dry air pump for carrying away the vapour, and the other a wet pump for removing the water. The former is connected to the top of the condenser (the vapour preferably being cooled before admission to the pump), and the latter to the lowest part of the condenser. It is somewhat doubtful, however, whether the advantage gained outweighs the additional cost and complication involved.

Vacuum Augmenter.—In order to assist the air pumps to deal with vapour Mr. Parsons has introduced what he calls a vacuum augmenter. This is a small apparatus fitted between the condenser and the air pump, and through which a jet of live steam is blown; this jet draws out considerable quantities of vapour from the pipe to which it is fitted, compresses it, and delivers it to the air pumps.

In striving for a high vacuum there is one point which must not be overlooked, it is this—It is quite useless to have a very high vacuum in the condenser if, owing to the want of area of the exhaust ports, a correspondingly high vacuum is not obtained in the cylinder of the engine. The reader will see from the last column of Table XVI. how enormous is the volume of a pound of steam at very low pressure (or high vacuum), and will realise how difficult it must be to get such a volume of steam out of a cylinder through its ports.

In Lancashire it has been found from experience that a vacuum of 26 inches gives the most economical results. The explanation doubtless is that, if a higher vacuum is obtained in the condenser, the exhaust steam must be cooled down to a much greater extent, and the temperature of the condensed steam which is fed into the boiler is correspondingly reduced. This reduced temperature of the boiler feed probably neutralises the gain due to a slightly better vacuum in the engine.

Corrosion of Condenser Tubes.—The galvanic action set up by Muntz metal tubes, brass tube plates, and the cast-iron shell tends to make the latter corrode. This in itself is not very harmful if the shell is made fairly thick in the first place, but if a piece of rusty iron lodges in one of the tubes it quickly corrodes its way right through. To overcome this source of trouble, Mr. Edwards advocates increasing the speed at which the water travels through the tubes, so as to sweep them more effectually. The usual speed is about 300 feet per minute; by greatly increasing this velocity Mr. Edwards claims that he has effected an improvement in the life of the tubes. Coating the inside of the condenser body with a wash of cement has proved very useful in preventing corrosion.

Edwards' Air Pump.—A pump which has come very largely into use during the last few years is Edwards' air pump, as shown by Figs. 67 and 68. This air pump differs from those of the older pattern, in that it has no bucket valves or foot valves,

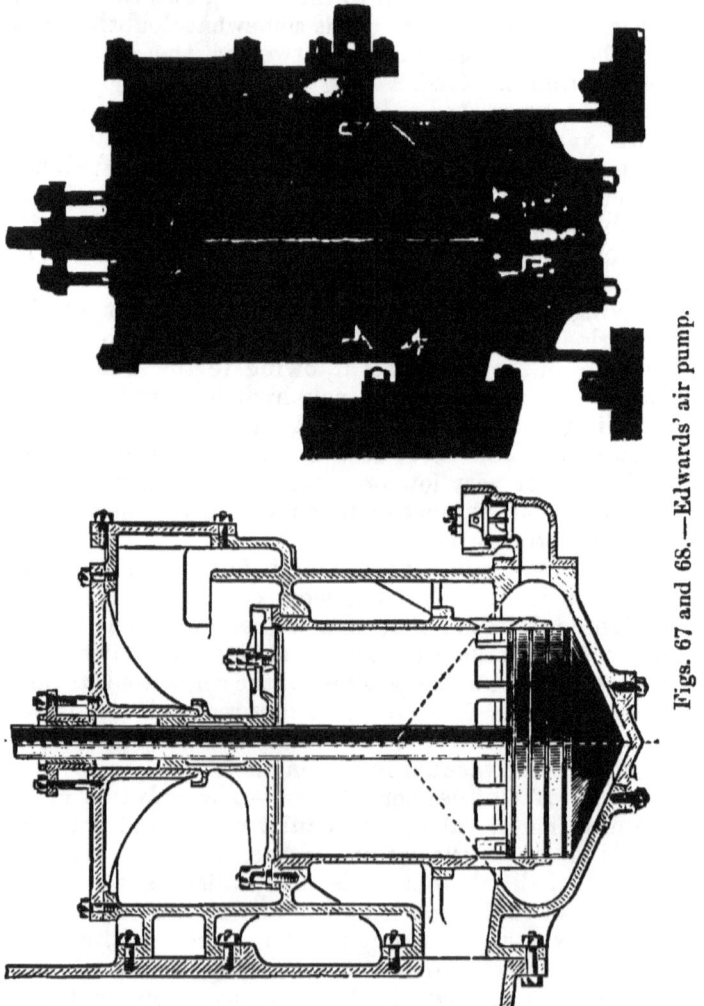

Figs. 67 and 68.—Edwards' air pump.

and runs at a considerably higher speed than was formerly considered practicable.

The pump is placed below the level of the condenser, so that the condensed steam flows continuously by gravity from the condenser to the base of the pump. On the downward stroke a vacuum is created between the bucket and the head valves, and at the bottom of the stroke the bucket uncovers a row of ports, so that the vapour from the condenser enters the barrel. By its conical shape the bucket projects the water which was lying in the base of the pump through the ports into the pump barrel; the water is projected with considerable force, and it also entrains a certain amount of vapour. On the up stroke the bucket closes the ports and sweeps out the water and vapour through the head valves.

This pump has many advantages; in the first place, the bucket valves and foot valves, which, in the old form of pump, were necessarily rather inaccessible, are eliminated. In the second place, when the bucket uncovers the ports the vapour has free entry to the barrel, whereas, in the old form of pump, a certain pressure in the condenser was necessary to open the valves. In the third place, owing to the high speed at which it is possible to run the pump, it deals with small quantities of water at a time, and runs very smoothly and with freedom from shock. There is a large inspection door, shown at the top right-hand side of the illustration, which gives access to the head valves.

Evaporative Condenser.—In cases where water for condensing purposes is scarce, or has to be paid for, and the amount of steam to be condensed is comparatively small, an evaporative condenser is sometimes employed. This form of condenser consists of a range of pipes having external gills, through which pipes the steam is passed. Water is allowed to trickle on to them, and the water is evaporated, thus extracting a considerable amount of heat from the pipes. By this system, instead of 30 to 40 times the amount of feed-water being required for condensing purposes, an amount of water equivalent only to the feed is required. The remarks in connection with power brakes in Chap. VI. will make the reason for this clear. The objection to this form of condenser is that it takes up a good deal of room, and the clouds of steam arising from it are undesirable in a town.

The Ejector Condenser.—In this condenser the exhaust steam and the cooling water are mixed together, and no air pump is required. The principle upon which the ejector condenser works is somewhat similar to that of the injector previously described, but, instead of water being fed into a boiler, the cooling water and condensed steam are discharged against atmospheric pressure into a hot well. Unlike the injector, the nozzle through which the steam passes is perforated with a large number of openings through which the water comes in contact with the exhaust steam and condenses it.

The ejector condenser to be really reliable should be supplied with cooling water from a tank 20 feet above it, or under a pressure of about 10 lbs. per square inch if supplied direct from a centrifugal pump. The advantage of this form of condenser is that it will discharge the condensed steam against atmospheric pressure while maintaining a vacuum of 24 or 26 inches in the engine without the aid of an air pump.

The ejector form of condenser is therefore very inexpensive, and no power is expended in driving an air pump. Such a condenser is, however, a little extravagant in cooling water; about 40 times the weight of the steam condensed is required, and the same quantity of water is required at light as at full loads. A non-return valve is invariably fitted between the engine and the condenser, but, even with this safeguard, there is a certain element of risk—viz., that of the water finding its way into the engine.

The danger can be avoided by carrying the exhaust pipe from the engine about 32 feet upwards and down again to the condenser, as water will not rise to this height under the vacuum formed. If, however, the pipes are carried up to such a height, a plain jet condenser of the form shown by Fig. 63 may be used. This would then be called a barometric condenser.

Barometric Condenser.—If a condenser, as shown by Fig. 63, is placed 35 feet above the level of the water into which it discharges, it will free itself of water by means of gravity. The pressure per square inch at the bottom of a column of water is ·433 lb. for every foot in height, so that a column of water 34 feet high exerts a pressure of 14·7 lbs. per square inch. Even if a perfect vacuum were formed in the condenser, the atmospheric pressure outside would not force the water to a height greater than this. If, then, the jet condenser condenses the steam at, say, 35 feet above the level of the discharge, the water will flow away by gravity, and the vacuum will still be maintained.

It has been found that the drops of injection water entrain any vapour, and that the velocity at which the water descends the pipe is sufficiently great to carry the bubbles down with it.

With this form of condenser no air pump is required, merely a centrifugal pump to assist in raising the cooling water to the required height. The amount of cooling water required is the same as in a jet condenser fixed in the ordinary way—viz., about 25 times the weight of the steam condensed.

Cooling Towers.—The chief difficulty in connection with the use of condensing plant in large towns has been the question of water supply. We have seen that the water required to condense the steam in a surface condenser is 40 or more times the weight of the steam used, so that in a large power installation the amount of cooling water required is very great indeed.

In the case of Lancashire cotton mills which are not on the banks of a canal, it has been customary to construct a fairly large reservoir of water, known as a lodge, from which the cooling water is drawn, and to which it is returned after passing through the condenser. The surface of the water being exposed to the atmosphere gives up a certain proportion of the heat extracted from the steam, and although towards the end of the day the temperature of the water, especially in hot weather, becomes rather high, yet the system is found to answer fairly well.

This method of cooling the water, although suitable in the case of a mill where the horse-power rarely exceeds 1,000 or 1,200, is not suitable, on account of the large size of reservoir required and the expense of the land needed, for a large power installation where many thousands of horse-power are developed.

The plan now usually adopted is to employ cooling towers. These towers, which range from 40 to 80 feet high, are filled with some material suitable for breaking up a mass of water and exposing as much of the surface as possible to the atmosphere. In one make of tower a large number of short earthenware pipes are used; they are stood up end to end, but the openings of the pipes do not come exactly over one another, thus the downward stream of water is continually broken up. The water from the condenser is pumped up to the top of the tower and trickles down the sides of the pipes, while a current of air rises up and meets it and extracts a good deal of heat from the water.

In another make of tower, galvanised wire and timber slats are used. Originally fans were used to send a current of air up the inside of the towers, but it has been found that in many cases the difference between the temperature of the air inside and outside the tower is quite sufficient to cause a good draught of air, and that a fan can be dispensed with.

The tower stands over a small tank or reservoir formed of concrete, into which the water is allowed to fall after passing down the tower. In large installations several towers are used.

A question which may occur to the student is—Why not dispense with water altogether and use air as the cooling medium in the first place? The reply is, that it is not possible to do so, as the specific heat of air is too low, and the volume of air required would be excessive. As nothing has yet been said about the specific of heat of substances, a little digression must be made.

Specific Heat is the amount of heat required to raise 1 lb.

of the substance through 1° F. Thus 1 British thermal unit will raise 1 lb. of water at its greatest density through 1° F. The specific heat of water is therefore said to be 1, and is generally adopted as the standard of comparison.

The specific heat of air is ·238 and of cast iron ·13, so that the specific heat of both air and cast iron is less than that of water; or, in other words, less heat is required to warm 1 lb. of either by 1° than is required to warm the same weight (not volume) of water. Conversely, a gas, such as air, the specific heat of which is low, is less capable of abstracting heat from another substance or gas, with a given rise of temperature, than a body the specific heat of which is high, such as water. To return to the condensing question.

The specific heat of air is ·238, so that if air were of the same weight as water, 4·2 times more air than water would be required for condensing. A cubic foot of air at 32° F., however, weighs only ·08 lb., while a cubic foot of water weighs 62·4 lbs., so that water is 780 times heavier than air. We should, therefore, require a volume of air 780 × 4·2, or 3,270 times greater than that of water, to obtain the same cooling effect. It would be out of the question to deal with such an enormous volume of air.

While upon the subject of specific heat, a few words on the subject of the specific heat of steam may be said, as this is of practical interest when dealing with superheated steam. The specific heat of steam was originally found by Regnault to be about ·48, and this figure is usually taken to be correct, although doubts have often been expressed as to its accuracy. Some experiments recently made at the Munich Technical School have thrown a good deal of light upon the subject. From the curves reproduced in *Engineering*, vol. lxxxiii., p. 227, it would appear that the specific heat of saturated steam varies from ·45 at atmospheric pressure to ·62 at 142·24 lbs. pressure, but as soon as the steam begins to get superheated the specific heat falls, until at 500° F. the specific heat varies from ·46 to ·5, according to the pressure, while at 600° F. the specific heat varies from ·475 to ·498.

CHAPTER X.

THE STEAM TURBINE.

THE great success which the steam turbine has achieved during the last few years seems to render it probable that before many years are past it will largely, if not entirely, supersede reciprocating engines for marine work and for driving electric generators. Instead of huge engines having big pistons, with heavy piston- and connecting-rods moving up and down and turning a crank, we have in a turbine a revolving drum receiving its motion direct from the steam.

The credit for this transformation is almost entirely due to the Hon. C. A. Parsons, who first believed in the possibility of constructing a turbine to give considerable power without an excessive consumption of steam. The first Parsons turbine of about 10 H.P. was constructed in 1885, and the first comparatively large turbine in 1890; and although during the next few years a fair number of turbines were made and supplied, their introduction was a stiff uphill fight. The earlier turbine undoubtedly used more steam than a good reciprocating engine, and the makers of such engines made the most of the fact; the turbine was referred to as a steam eater, and many were the jokes made as to its capacity in this respect. It was not until the Parsons turbine was worked in connection with a condenser that it was able to compete on terms of equality, as regards consumption of steam, with a reciprocating engine. When the figures obtained at the first authentic condensing turbine trial were published they were received with a certain amount of incredulity, but from that date the advance of the turbine into general favour has been steady and continuous.

The Parsons turbine is of the parallel flow, reaction * type— *i.e.*, one in which the steam flows through the turbine in a direction parallel with its axis, as shown by Fig. 69. The illustration shows a Parsons-Willans turbine (the difference between this and the original Parsons turbine is explained later) without bearings or governor. The blades in Fig. 69 are shown diagrammatically—*i.e.*, a row of blades is represented by one line; they are shown in detail by Figs. 70 and 72. The

* The meaning of the term reaction turbine is explained in the Chapter on Water Turbines.

rotating drum or rotor is provided with a large number of blades, and between each row of moving blades there is a row of fixed guide blades attached to the casing, as shown by Fig. 70.

Steam at high pressure is admitted at A, Fig. 69. It passes through the guide blades and impinges on a row of blades attached to the drum; it exerts a reactionary force on these blades, causes them to move and thus rotates the drum to which they are attached. After passing through the next row of guide blades the steam impinges on the succeeding row of blades, and so on, until the steam is fully expanded, when it passes away to the condenser.

It will be seen from Fig. 69 that at first the blades are short, they are also closely spaced; as the steam expands blades of a greater length and width, and more coarsely pitched, are used.

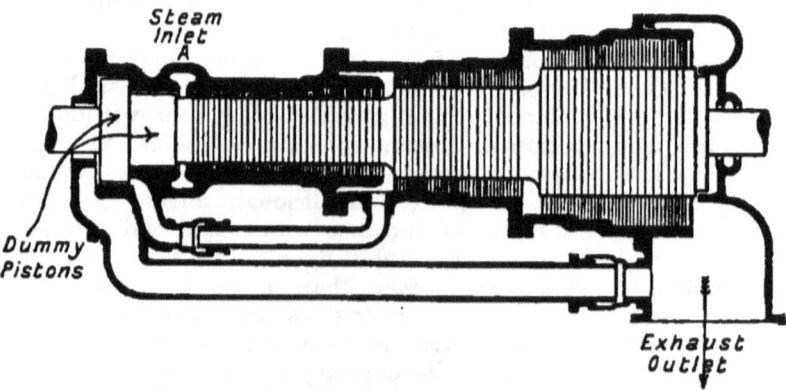

Fig. 69.—Willans-Parsons turbine.

The drum also is increased in diameter; this allows of a larger number of blades being used, and the somewhat weaker turning effort of the steam on them is made up for by their increased distance from the centre of the drum, or, in other words, by their increased speed. By the time the steam has reached the end of the turbine it has, by a long series of steps, fallen in pressure and has imparted a large portion of its energy to the rotating drum.

In order to prevent end thrust the drum is enlarged at the high-pressure end of the turbine; these enlargements are called dummy pistons, and are provided with baffle rings, as shown by Fig. 71. The turbine is usually arranged so that high-pressure steam can be admitted by a pipe or passage to the low-pressure end; this enables additional power to be obtained for short periods, but of course uneconomically. The pipes shown underneath the turbine are for balancing purposes. The small pipe allows steam at a pressure corresponding with that in the middle of the turbine to press against the face of the dummy piston,

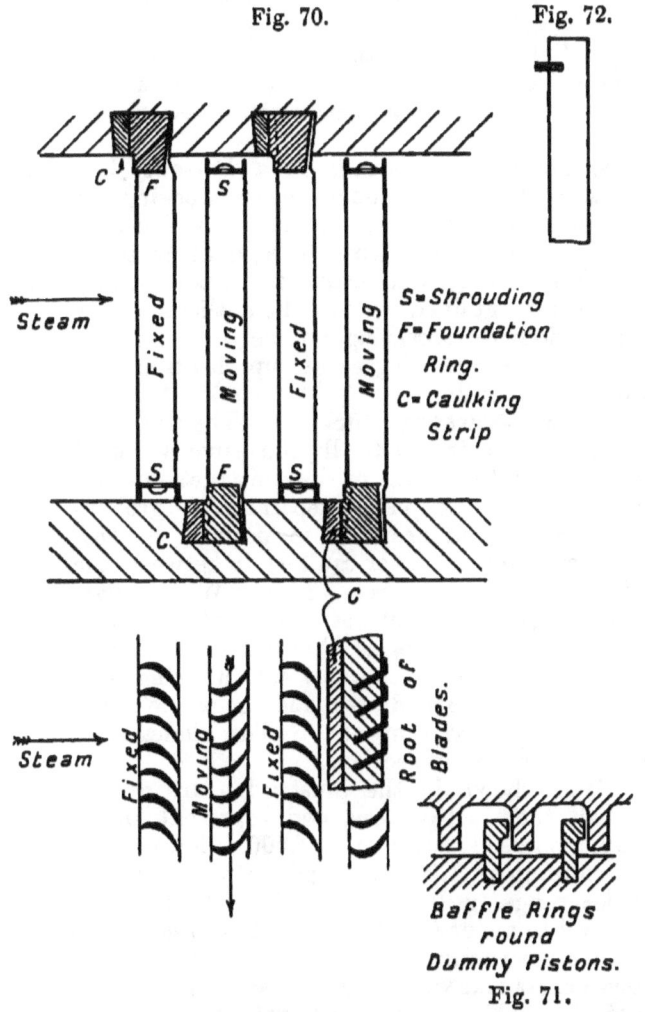

Figs. 70 to 72.—Turbine blading.

while the large pipe places the back of the piston in communication with the condenser.

In the Parsons turbine the governor is driven from the main shaft by a worm and wheel, and is arranged so that the steam is admitted intermittently or in puffs; when the turbine is lightly loaded the interval between the puffs is longer than when fully loaded. With a full load on the turbine the puffs are almost continuous. The effect of this arrangement is that the turbine is always supplied with steam at high pressure even when working at light loads. If the governor worked on the ordinary throttling system the turbine, at light loads, would be supplied with steam at a pressure considerably less than that of the boiler.

The consumption of steam in a Parsons' 3,500 kilowatt turbine, say 5,100 brake horse-power, working with 200 lbs. steam pressure, and 121° of superheat, has been as low as 13·2 lbs. per kilowatt hour. This is equivalent to 9·85 lbs. per electrical horse-power, and about 9 lbs. per brake horse-power. The best result obtained, within the author's knowledge, with a triple-expansion reciprocating engine working with the same pressure and the same degree of superheat, has been 11·9 lbs. per brake horse-power.

The Willans-Parsons Turbine.—The principle upon which this turbine works is practically the same as the Parsons, but there are certain differences in the construction. In the Parsons turbine the blades are placed separately in grooves cut in the rotor and casing and are held in position by small pieces of bronze which are wedged or caulked in between the blades, the ends of the blades being free. In the Willans-Parsons turbine the outer ends of the blades are riveted into an encircling U-shaped ring of bronze, as shown by Fig. 70. The root of each blade is inserted into a saw cut in a ring, one side of the blade at the root is turned over, and the whole is wedged tightly in its groove by a caulking strip. The blades with their rings are put into position in sections.

The blades shown by the illustration are drawn to approximately half their actual size, and represent those used about half way along the rotor of a 1,500 kilowatt (2,200 B.H.P.) turbine running at 1,500 revs. per minute. In such a turbine the blades range in length from 1 or 1½ inches at the high-pressure end to about 6 inches at the low-pressure end. In large marine turbines running at a considerably lower speed the blades are much longer and wider. The blades at the low-pressure end of one of the turbines for an Atlantic liner, the "Mauretania," are about 23 inches long and 2 inches wide.

The U-shaped shrouding ring at the ends of the blades is used chiefly in the Willans-Parsons turbine (and by Messrs. Yarrow); in the original Parsons turbine the blades were strengthened by a ring let into the blades, and soldered to them by silver solder, as shown by Fig. 72, at the top right-hand corner of the illustration; if the blades were very long, two or more rings were used.

The shrouding, besides strengthening the blades, has two other advantages—viz., should any whipping of the shaft or hogging of the casing take place and cause the blades to touch the casing, the shrouding will merely rub against the latter; the shrouding will certainly wear away, but the blades will not be stripped away, as is the case if the ends are unprotected. The other advantage of shrouding the ends and having a ring round the roots of the blades, is that the flow of steam past the ends is to a certain extent checked, and the loss due to leakage past the ends is probably less than with blades having free ends.

The consumption of steam is approximately the same as in the Parsons turbine.

In some of the Willans-Parsons turbines the upper portion of the casing is hinged, so that there is no danger of damaging the blading when the upper half is being opened up for inspection or during replacement. All pipes and connections are attached to the lower half of the casing. The Willans-Parsons turbine is governed by throttling, and not by admitting steam intermittently.

The Brush-Parsons Turbine.—This turbine is very similar to the Parsons; it differs only in minor points of construction; for instance, the blades are constructed with a strengthening ring let into their ends, instead of into one side as in the Parsons. A special centrifugal water gland is used for packing the shaft at the ends of the turbine.

The speeds at which the Parsons, Willans-Parsons, and Brush-Parsons turbines are usually designed to run, when driving electric generators, are as follows:—

TABLE XVII.

Electrical Output.	Brake H.P.	Revolutions per Minute.
500 Kilowatts.	730	1,500 to 3,000
750 ,,	1,100	1,500 ,, 3,000
1,000 ,,	1,450	1,500 ,, 1,800
1,500 ,,	2,200	1,500 ,, 1,800
3,000 ,,	4,400	750 ,, 1,500
5,000 ,,	7,300	750 ,, 1,000

For marine work the speed of the turbine is much less; for instance, the H.P. and L.P. turbines in the Cunard Liner

"Lusitania," which collectively give about 64,000 H.P., run at about 188 revs. per minute.

When the turbine is coupled direct to an alternator, the exact speed is determined by the periodicity of the latter. The meaning of the periodicity of an alternator is given in the electrical chapter. The peripheral speed of the blades in turbines of the Parsons type does not usually exceed 300 feet per second.

The De Laval Steam Turbine.—This turbine, which was invented by a Swedish engineer, works on a principle different from that of the Parsons. In the latter the steam is, as we have seen, expanded from its highest to its lowest pressure through a long succession of steps, each row of blades absorbing a small part of the energy of the steam.

In the De Laval turbine, the steam is expanded in one step from its highest to its lowest pressure. This expansion is carried out in a nozzle. It has been proved by experiment that, although steam loses its pressure if expanded from a small volume to a larger volume, it does not lose its temperature or energy, provided it does no external work during such expansion.*

In the De Laval turbine this fact is made use of, so that, instead of having a small volume of steam issuing from a nozzle at a high pressure (in which case much of the steam would expand in the air after striking the buckets, and its energy be lost), the nozzles are constructed so that the steam is expanded in them before leaving the orifice. We have, therefore, a very large volume of steam at low pressure travelling at high speed; by allowing the steam to impinge on buckets, also travelling at a high speed, we utilise a very large part of its energy. A turbine working on this principle is called an impulse turbine. The meaning of the term impulse turbine is explained in the chapter dealing with water turbines.

Fig. 73 shows the ring of buckets and four nozzles of the De Laval turbine; the illustration shows clearly the action of the steam on the blades. In order to take the greatest advantage of the energy of the steam, the ring of buckets must travel at a very high speed indeed; the peripheral speed of the buckets should be 47 per cent. of the velocity of the steam, so that, if the steam leaves the nozzle at a speed of 4,000 feet per second, the

---

* Joule's law is—"When a gas expands without doing external work, and without taking in or giving out heat, its temperature does not change." To prove this rule, Joule connected a vessel containing compressed gas with another vessel that was empty by means of a pipe with a closed stop cock. Both vessels were immersed in a tub of water, and were allowed to assume a uniform temperature. Then the stop cock was opened, the gas expanded without doing external work, and finally the temperature of the water in the tub was found to have undergone no change.

peripheral speed of the buckets should be 1,880 feet per second; but for practical reasons the speed is considerably less.

In the case of a 300 H.P. turbine the outside diameter of the wheel is $31\frac{1}{2}$ inches; it runs at 10,600 revs. per minute, giving a peripheral speed of about 1,457 feet per second or 87,420 feet per minute. With smaller turbines the peripheral speed is less.

Fig. 73.—Bucket wheel of De Laval turbine.

To enable a wheel to run at this enormous speed several interesting methods of construction have been adopted. The wheel is of the disc form, and in the larger turbines it is solid throughout; there is not even a hole through the boss, the shaft being bolted to the wheel by flanges on each side; the blades or buckets are dovetailed into the rim. The centrifugal force on these buckets is considerable; a bucket weighing $\frac{1}{2}$ oz.

if rotated at a radius of 1·25 feet, will develop a centrifugal force of about 13·4 cwts., when run at a speed of 10,600 revs. per minute. The weight of a bucket of a 300 H.P. turbine is a little heavier than ½ oz. The ends of the buckets form what is practically an encircling ring; it would be impossible to surround the buckets (of the larger turbines) with an encircling ring of steel, as is sometimes stated to be the case. If the reader will work out the stress in such a ring due to centrifugal force (the rule is given in Chapter VII.), he will find that it amounts to about 95 tons per square inch, or sufficient to burst any steel ring. Each bucket is firmly dovetailed into the disc, and has only to withstand the stress due to its own centrifugal force.

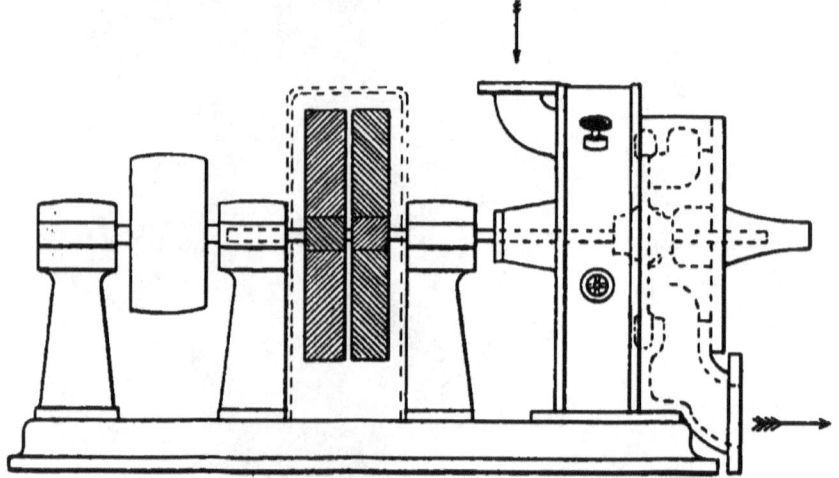

Fig. 74.—De Laval turbine with gearing.

The shaft carrying the bucket wheel is of small size; that of a 300 B.H.P. turbine is only $1\frac{5}{16}$ inches diameter, the reason for this being that it is impossible to insure that the centre of the shaft shall be absolutely in the centre of gravity of the wheel, or, in other words, that the wheel shall be in perfect balance. If a very stiff shaft were used the effect of a slight want of balance would, at the very high speeds employed, be to cause excessive vibration. By employing a light and, therefore, somewhat flexible shaft, the shaft is enabled to spring somewhat; when the wheel reaches about $\frac{1}{6}$ or $\frac{1}{8}$ of its full speed, it settles itself on a new centre so as to run smoothly and free from vibration.

The speed of the De Laval turbine, which ranges from 30,000 revs. per minute in the 5 H.P. turbine, to 10,600 revs. in the

300 B.H.P. turbine, is too high to enable it to drive even a dynamo if directly coupled to it. The speed is therefore reduced by double helical gearing, as shown by Fig. 74. The two halves of the helical teeth are slightly separated, so that they can be machine cut. The pinion is made in one piece with the shaft, and the linear velocity of the teeth is about 1,000 feet per second. The helical wheels are enclosed in a casing, as shown by dotted lines. The illustration shows the turbine driving a belt pulley, but a dynamo, pump, or fan, can be placed in the position occupied by the pulley.

The turbine is provided with several nozzles, some of which are usually closed when working with high steam pressures and condensing; some of the nozzles are also closed when the turbine is working lightly loaded.

A consumption of 18·9 lbs. per kilowatt hour, working with 193 lbs. pressure, and with 60° of superheat, has been recorded. The De Laval turbine is made in this country by Messrs. Greenwood & Batley of Leeds.

The Curtis turbine, made by the British Thomson-Houston Co., is of the "Impulse" type, but is designed so that an extremely high rotative speed is not necessary in order to obtain good results.

In the Curtis turbine, as in the Laval, the steam is expanded in nozzles before striking the buckets, but not to the same extent. The nozzles are designed so that the speed of the steam issuing from them shall be about 2,000 feet per second. After the steam has passed through one row of buckets its course is altered by stationary buckets or blades, so that the steam shall impinge to the best advantage on a second row of moving buckets. After the steam has passed through this second row of buckets, which completes one stage, it is again passed through nozzles, and goes through a course similar to that just described—viz., moving buckets, stationary buckets, and moving buckets.

Fig. 75* shows the arrangement of nozzles and buckets in a two stage turbine. The two rows of moving buckets are bolted to the upper and lower sides of one rotating disc.

Fig. 76 shows the general arrangement of the Curtis turbine driving a dynamo, the latter being placed immediately above the turbine.

It will be seen from Fig. 76 that the Curtis turbine, in its larger sizes, has, unlike the Parsons and the De Laval, a vertical

* Fig. 75 causes a curious optical illusion, each row of buckets apparently varying in width. Measurement by means of a pair of dividers will show that this is not so.

shaft. This shaft, which carries the whole of the revolving portions of the turbine and dynamo, is supported by a footstep bearing. The latter is made in halves; the surfaces are kept apart by a film of oil or water supplied under pressure. The lubricant, after leaving the footstep bearing, passes upwards, and lubricates a guide bearing which keeps the shaft central. The upper guide bearings at A and B are also lubricated by oil supplied under pressure; after use, the oil passes to a tank, and is used over and over again.

The governing of the Curtis turbine is effected by opening or closing the valves controlling the nozzles, as shown by Fig. 75, the amount of steam admitted to the turbine being proportional to the load; the steam pressure is not reduced by throttling.

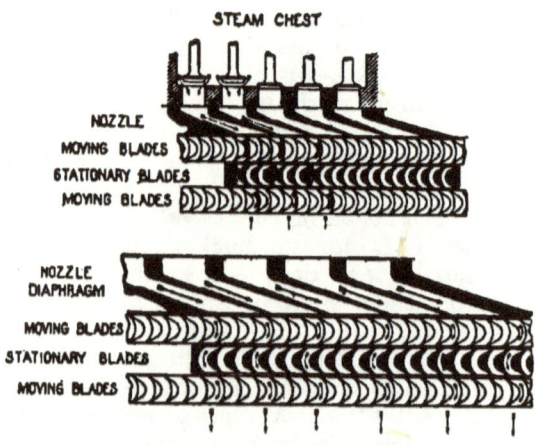

Fig. 75.—Curtis turbine blading.

The buckets of the Curtis turbine are cut out of a solid ring which is bolted to a disc. A thin encircling ring of steel surrounds the outer ends of the buckets.

The clearances of the Curtis turbine, unlike those of the Parsons, can be adjusted to a nicety by means of a strong screw at the bottom of the footstep bearing. In the larger turbines this screw is worked by a worm and wheel. The buckets being shrouded, if any rubbing should occur the shrouding will wear, but the buckets will not be destroyed.

The lower part of the Curtis turbine usually contains the surface condenser. The Curtis turbine occupies much less floor space than one of the Parsons type, and the construction of the

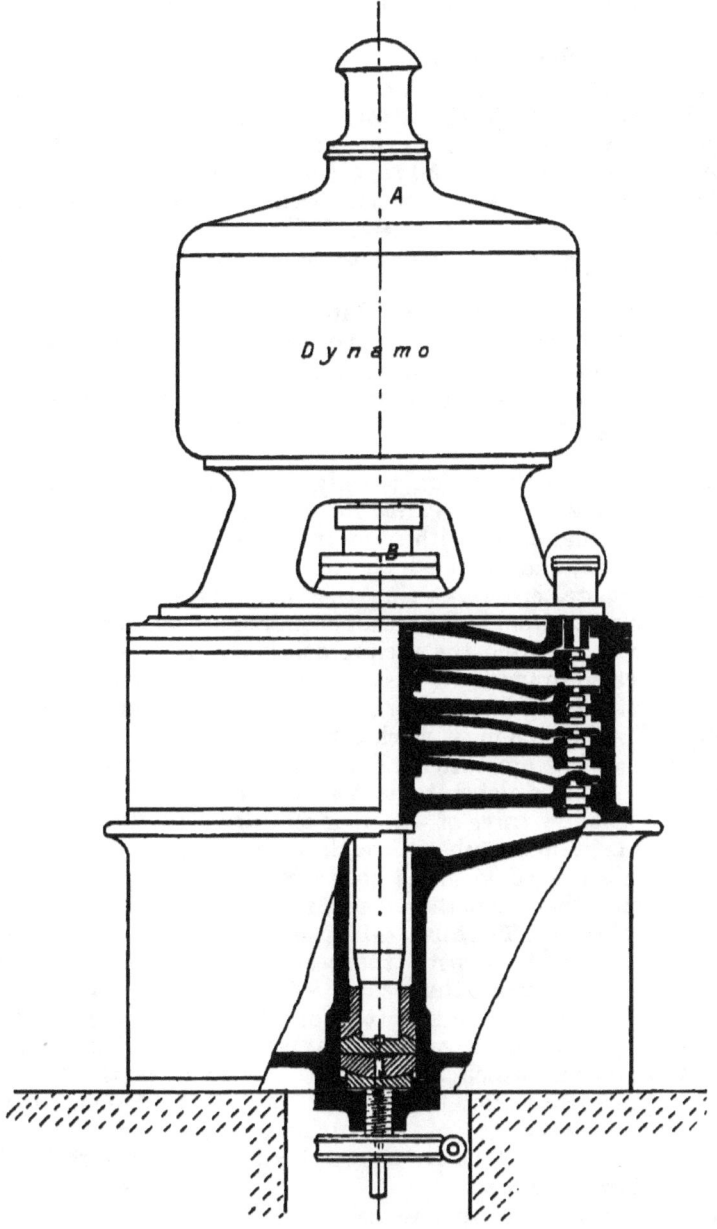

Fig 76.—Curtis turbine.

buckets, cut out of the solid as they are and shrouded, is a good feature.

The consumption of steam in Curtis turbines usually employed for driving generators up to 3,000 kilowatts, working with superheated steam and with a good vacuum, is frequently about 16·5 lbs. per kilowatt hour; but in the case of a 10,000 kilowatt five-stage turbine, running at 750 revs. per minute, a consumption as low as 12·9 lbs. per kilowatt has been recorded. The steam pressure was 176 lbs.; superheat, 147°; vacuum, 29·47 inches. This turbine was one of four erected at the Fisk Generating Station, Chicago.

The Rateau Turbine.—The Rateau turbine, made by Messrs. Fraser & Chalmers, is similar in principle to the Curtis, but the shaft is horizontal instead of vertical. This turbine has been employed to a considerable extent on the Continent, frequently being driven by the exhaust steam from reciprocating engines.

In cases where the steam is only supplied intermittently, a heat accumulator is used. This, in its simplest form, consists of an old boiler shell filled with scrap iron, into which the exhaust steam is taken, and from which the turbine draws its supply. The exhaust steam, on entering the shell, gives up some of its heat to the scrap iron, and partially condenses; when the supply of exhaust steam from the reciprocating engine ceases, and the turbine goes on drawing its steam, the pressure in the shell falls, and the water lying in it turns into steam, and so keeps up the supply. The turbine must work condensing; otherwise, steam at the somewhat low pressure available would be of comparatively little use. A more elaborate heat accumulator consists of a shell containing shallow trays of water, or of a drum partly filled with water, and containing tubes through which the steam passes, the tubes being arranged so that a good circulation of water is maintained around them, but the principle is the same.

Westinghouse Turbine.—In this turbine the impulse principle is combined with the reaction or Parsons system. Steam is admitted at the centre of the turbine, and flows outwards towards the ends. It impinges first on two impulse wheels, as in the De Laval turbine; after passing the impulse wheels the steam, which has dropped to about one-third of its original pressure, passes through blading of the Parsons type, and leaves at each end of the turbine. By this arrangement dummy balancing pistons are not required, and the overall length of the turbine is reduced.

A somewhat similar turbine has recently been constructed by Messrs. Melms, Pfenninger & Sankey, but in it the steam is admitted at one end, strikes an impulse wheel, and then passes through blading of the Parsons type to the other end, where it goes to the condenser. With a 500-kilowatt turbine constructed on this principle, a consumption of 17·2 lbs. per kilowatt hour has been recorded.

The Zoelly Turbine works on the same principle as the Curtis; but, like the Rateau, the shaft is horizontal. There are usually ten rotary wheels having buckets round the periphery; these wheels are placed in two separate casings; the shaft runs right through the casings, and is supported by a bearing placed between them. One casing is the high-pressure portion of the turbine, the other the low-pressure portion. The steam, after passing through the high-pressure portion, is conveyed by a pipe to the low-pressure portion. If the reader will turn the illustration of the Curtis turbine up sideways, so that the shaft is horizontal, and imagine that the dynamo shown is the high-pressure portion of the turbine, he will have a good idea as to the general appearance of the Zoelly turbine.

In this turbine the blades are made of nickel steel, and the section decreases from the roots to the tips.

General Remarks on the Steam Turbine.—It has already been said that turbines of the Parsons type when coupled to dynamos have given results, as regards steam consumption, which have not been equalled by reciprocating engines. It may, therefore, be well to look for the reasons which enable such economy to be obtained. In the first place there is but little initial condensation, as the steam after doing its work passes away at the end of the turbine farthest from that at which it was admitted. Condensation in the turbine can also be reduced by superheating the steam to a high degree, and it must be remembered that a turbine can be supplied with superheated steam with much less risk of injury than a reciprocating engine, as in the former there are no rubbing surfaces in contact with the steam. The bearings require to be oiled, but these are outside the turbine proper, and the steam does not reach them.

Another point which conduces to economy is this—owing to the absence of internal friction, it is worth while to expand the steam to a point much further than would be useful in the case of a reciprocating engine. In a turbine, too, it is possible to provide a very large opening through which the steam can pass away to the condenser, as the reader will see by referring to Fig. 69. It is found that a difference of even half an inch in the vacuum makes an appreciable difference in the consumption of steam. The reader will have seen from the table given in the last chapter that the volume of steam increases very rapidly as the vacuum approaches a perfect one; in fact, the volume of steam with a 29-inch vacuum is double that of steam with a 26-inch vacuum, so that with the high vacuum we have double the volume of steam acting on the last blades of the turbine. The saving of steam due to the last few inches of vacuum is at the rate of between 4 and 5 per cent. per inch of vacuum.

The economy of a steam turbine would probably be still greater were it not for the necessity of having clearance between the ends of the moving blades and the casing, and between the ends of the fixed blades and the drum. Owing to the somewhat great length of the turbine, the casing has a tendency to "hog" under the temperature and pressure of the steam. The drum, too, has a tendency to "whip" owing to its length and high speed, and to the impossibility of making a perfect balance of the blades. To insure that the blades will not come in contact with the casing, it is necessary to give them a certain clearance, and this clearance allows a percentage of steam to pass without doing any effective work. In small turbines the clearance is proportionately greater than in turbines of large power, and it is due to this fact that turbines of less than 1,000 H.P. cannot at present compete successfully on the score of economy with triple-expansion reciprocating engines.

In spite of this clearance loss, the fact remains that turbines of large size (as already stated) are more economical than reciprocating engines. Apart from the economy attained with turbo-electric generating plant, it has been found that the coal used on the Channel steamers fitted with turbines is from 15 to 20 per cent. less than on those fitted with reciprocating engines.

Other points in favour of the turbine for marine work are these:—The weight of turbines and boilers for propelling a ship is about 5 per cent. less than that of reciprocating engines and boilers of the same power. In small vessels of the torpedo-destroyer class, the whole of the turbine can be placed below the water line, so that there is less risk of damage from gun fire than is the case with a reciprocating engine. For passenger steamers the absence of vibration is a great advantage. A good feature of the turbine for both land and marine work is the fact that no oil is required in the turbine itself; the condensed steam may therefore be pumped back into the boiler without the intervention of oil filters.

Both for land and marine work the turbine has proved itself

to be thoroughly reliable. An incident which was related to the author a good many years ago by the engineer concerned may not be considered out of place. Amongst the earlier turbines made by Messrs. Parsons were some constructed for one of the London electric light companies. These turbines were placed on the first floor of the electric generating station, while the ground floor was occupied by high-speed reciprocating engines. When a serious fire broke out one night, the greatest efforts were made to keep up a supply of current, and the engineer succeeded in keeping the turbine sets running after all the reciprocating sets had failed owing to the heat and water. The load which the turbine sets had to take up was enormous, but by throwing buckets of water on the dynamos to keep them cool, the turbine sets came through the ordeal triumphantly.

The reliability of the turbine, together with the fact that it may be over-loaded without serious risk and without a great falling off in economy, are points very greatly in its favour. In the case of electric generating stations, the saving effected in the cost of dynamos, and of buildings and foundations, to say nothing of the saving in oil and engine-room attendants, is very marked indeed.

The limitations of the turbine are chiefly those due to its high speed; the speed is too high to enable it to drive by belts or ropes without the intervention of gearing. If, however, the electrical method of transmitting power is adopted this objection does not hold good. One rolling mill has recently adopted this method of transmission; in this case steam turbines are used for driving dynamos, and the rolls for rolling steel rails are driven by motors. The arrangement is said to give entire satisfaction. A similar arrangement has recently been introduced into a cotton mill.

The steam turbine is somewhat handicapped by the fact that it does not reverse. On board ship this difficulty is overcome by having separate turbines for reversing. In marine work several turbines are usually employed for driving the ship; for instance in the "Carmania," an Atlantic liner, one high-pressure turbine drives a propeller amidships and two low-pressure turbines drive the port and starboard propellers. The reversing turbines also drive these shafts; when going ahead the reversing turbines remain connected to the condenser, steam being, of course, shut off.

Another fact which has somewhat retarded the introduction of the turbine is that it must condense if it is to complete successfully with a reciprocating engine. In large towns water must be paid for at a fairly high rate and this charge would be prohibitive if the cooling water were used only once and then thrown away; but the tendency now is for generating stations to be built on the outskirts of large towns where sufficient land is available to admit of cooling towers being erected. The water is cooled in these towers and is used over and over again.

On board ship the difficulty does not occur as sea water is available for cooling purposes.

CHAPTER XI.

ELECTRICAL CHAPTER.

It is almost essential at the present time that an engineer should have some knowledge of electrical matters, and while no attempt will be made to go at all deeply into the subject in the present book, yet a few words dealing with the main facts of practical work and with the relation of electrical units, such as the volt, ampere, and ohm, to mechanical units of work, such as horse-power, may be useful.

Production of the Electric Current.—In all cases where current is required in any quantity or at any useful pressure (electromotive force) it is produced by a dynamo or alternator, and the principle upon which these machines work will now be described.

In an ordinary horse-shoe magnet, such as may be purchased in a toy shop, there are two poles, one north, the other south; if the magnet is straightened out as in a mariner's compass the north pole (or north-seeking pole) will point to the north and the south pole to the south. Surrounding each pole is a magnetic field, and for the purpose of making calculations in designing dynamos the strength or weakness of a magnetic field is expressed by a number of imaginary magnetic lines of force per square inch of the magnet face. When the two poles of a magnet are brought together, as shown by Fig. 77, the lines of force are supposed to flow from the north to the south pole and constitute a magnetic field. The air is a bad conductor of these lines of force, and if the poles are wide apart a weak field results. Soft iron, on the other hand, is a very good conductor, and if a portion of the space between the poles of a magnet is filled with soft wrought iron, it assists the lines of force to flow, and the space left unbridged has a stronger field than would be the case without such partial bridging.

A horse-shoe magnet of the kind described is called a permanent magnet. A much more powerful magnet can be made by passing an electric current round and round a bar of soft iron or high permeability steel, and such electrically-excited magnets are used in practice, excepting for the very smallest class of dynamos, which are then usually known as magnetos.

If a wire is placed between the legs of the magnet, as shown by Fig. 77 (the black dot represents the wire in section), and is moved sharply downwards so that it cuts the lines of magnetic force, an electromotive force (E.M.F.) is induced in the wire, and if the ends of the wire are joined so that the circuit is completed, a current of electricity flows round it. The E.M.F., which may be looked upon as electrical pressure, is spoken of as so many volts; the amount of current which will flow through the wire (if the ends are joined) depends upon the E.M.F. and the resistance of the wire.

With a permanent steel magnet and single wire the E.M.F. induced would be too small to produce a serviceable current, and obviously it would be very inconvenient to move the wire rapidly up and down as shown. In actual practice an electrically-excited magnet or number of magnets are used, and the wire, or

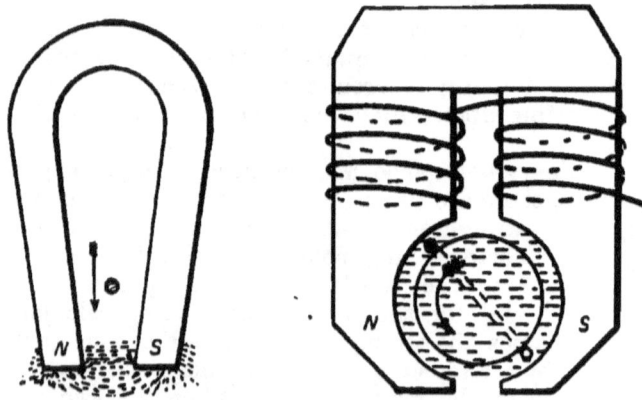

Fig. 77 and 77a.—Horse-shoe magnets.

wires, which have to cut the lines of force are placed on the outside of a drum, as shown by Fig. 77a. The drum is made of soft iron plates so as to facilitate the passage of the lines of force; by rotating the drum rapidly the wires on its outside pass through a strong magnetic field, or, in other words, cut a large number of lines of force flowing at right angles to the wires. Such a drum with wires is called an armature, although an armature need not necessarily take the form of a drum. It should be mentioned here that in order to produce E.M.F. in a wire the latter must pass from a weak field, or no field, into a strong field or *vice versâ;* or, if moved continuously in a field of the same strength, the speed must be varied, otherwise no E.M.F. will be obtained.

The amount of E.M.F. produced in a single wire on a drum, when rotated in a magnetic field, depends upon the strength of the field (or number of lines of force cut by the wire), and the rapidity with which the wire is moved. Assuming that one wire cuts sufficient lines of force at a speed sufficiently great to produce an E.M.F. of 1 volt, then, if the drum is wound so that there are 100 turns of wire round it, and it is rotated at the same speed as before, an E.M.F. of 100 volts will be produced, as the voltage in each turn of the wire adds itself to the voltage produced by the other turns. The wire must, of course, be insulated and wound in the right way.

We have said that the amount of current which will flow in a wire is dependent upon the E.M.F. and the resistance of the wire, or, expressed as a formula, $C = \dfrac{E}{R}$. The amount of current is expressed in amperes, and the resistance of a wire in ohms,* or parts of an ohm, so that the formula, translated into actual working terms, is amperes $= \dfrac{\text{volts}}{\text{ohms}}$. Thus, assuming the E.M.F. in a wire is 100 volts, and the resistance when the circuit is completed is 2 ohms, we shall have a current of 50 amperes flowing. If, on the other hand, we have an E.M.F. of 1 volt and a resistance of 2 ohms, we shall only have a current of half an ampere flowing.

To return to the dynamo. If a single wire is used, as shown in black by Fig. 77a, when the drum is rotated and the wire cuts the lines of force by descending through them on the left-hand side, the current will flow towards the spectator; when the wire cuts the lines of force on the right-hand side by rising through them, the current will flow away from the spectator.˜ Thus the current in the wire is reversed during every revolution of the drum, and is called an alternating current. A dynamo arranged with the wires or conductors arranged to give this effect is called an alternating current generator, or alternator.

If, instead of a single wire on one side of the drum only, the wire is prolonged and wound round the drum, as shown by dotted lines in Fig. 77a, it will at once be seen that the current flowing towards the spectator on the left-hand side and away from him on the right-hand side will produce a continuous current in the encircling wire, but the direction of the current will be reversed once during every revolution. If, however, the current flowing round the armature can always be tapped or

* Table XVIII., giving the resistance of wires of various sections in ohms, will be found at the end of this chapter.

drawn off at the point where its E.M.F. is highest, and where the current is flowing, say towards the spectator, and returned to the armature at a point where the current is flowing away from the spectator, a continuous or direct current will be obtained, and will be available for use outside the dynamo. This tapping of the current at the right place is effected by means of a commutator. The commutator consists of a number of segments of copper insulated from each other, and from the shaft; to each one of these segments one end of a wire, wound round the armature, is connected, the other end being connected to the next segment. In actual practice the wire is taken a good many times round the drum before its ends are connected to the commutator segment. Brushes consisting of copper gauze pressed tightly together, or of carbon, are used for collecting the current. One brush or set of brushes is placed over, and pressed by a spring upon, the segment of the commutator where the current is at the greatest E.M.F., and where it is flowing in the right direction; another set of brushes returns the current to a segment at the opposite of the commutator, thus completing the circuit. A dynamo provided with such a commutator is called a continuous-current or direct-current generator. If a large number of poles are used, then a set of brushes is provided to collect the current from the commutator at each pole.

We have seen that the volt is the unit of E.M.F., the ampere the unit of current, while the ohm is the unit of resistance to the passing of the current. One ampere of current at a pressure of 1 volt is called 1 watt. Thus 5 amperes at a pressure of 2 volts = 10 watts, or 2 amperes at a pressure of 5 volts are also 10 watts. A thousand watts are called 1 kilowatt, and dynamos are usually spoken of as giving so many kilowatts. Thus a dynamo which gives an output of 100 amperes at 100 volts is a 10-kilowatt dynamo. A dynamo which gives 1,200 amperes at 100 volts is a 120-kilowatt dynamo; while a dynamo which gives 3,000 amperes at 500 volts is a 1,500-kilowatt machine.

Electrical current cannot be produced without the expenditure of energy, and it may be useful at this stage to see what relation the electrical units bear to a mechanical horse-power. The relation is this—746 watts = 1 electrical horse-power. An electrical horse-power is, theoretically, the equivalent of 1 mechanical horse-power, and if there were no losses in a dynamo, 1 brake horse-power transmitted to it would produce 746 watts. But, as a fact, there is, and must be, a certain loss of energy in every dynamo. In the first place, there is the resistance of the wires or coils carrying the current round the armature, and, in the second place, there is the loss of that portion of the current which is used for exciting the magnets. There are also the losses due to mechanical friction and windage.

In a well-designed dynamo of, say, 200 kilowatts and upwards, these losses may not exceed 6 per cent., so that for every 100 brake horse-power transmitted to the dynamo, the latter will give an output of 94 E.H.P.; in this case the efficiency of the dynamo is said to be 94 per cent. In small dynamos such a high efficiency is not obtained, while in very large dynamos the efficiency may be greater.

A continuous-current dynamo can be used either for generating electricity or it can be used for converting electrical energy back into mechanical work; when a dynamo is used for the latter purpose, it is called a motor. The electrical losses in a motor are the same as in a dynamo, but in a small motor of, say, 10 H.P., one would not expect an efficiency higher than 88 or 90 per cent.

It may be of interest to see how far an electrical horse-power will go in the way of current for lighting purposes. A 16-candle-power 200 to 250 volt incandescent (glow lamp) requires, when new, about 60 watts per hour, so that $12\frac{1}{2}$ lamps will absorb about 1 E.H.P (746 watts); 124 lamps will absorb 10 E.H.P.; and 1,244 lamps 100 E.H.P. The student must not, however, fall into the mistake of thinking that a steam engine indicating 100 H.P. is capable of driving a dynamo when supplying current for 1,244 lamps, as the losses already referred to must be taken into account. In the first place, the loss due to friction in the engine itself will probably reduce the available power to 90 B.H.P. Then, assuming the efficiency of the dynamo is 92 per cent., we shall have only 92 per cent. of 90 B.H.P., or 82·8 E.H.P., to dispose of. If another 5 per cent. (say 4·2 H.P.) is lost through the resistance of the wires or leads conveying the current to the lamps, we have only 78·6 E.H.P., or 58,635 watts, to dispose of. Assuming each lamp requires 60 watts, then a 100 I.H.P. engine coupled to a dynamo will supply current for 977 lamps. In practice it is usually reckoned that 1 I.H.P. is required for 8 to 10 16-candle-power incandescent lamps.

Arc lamps, or those in which the light is caused by the current tearing away and making incandescent small particles of carbon in jumping from one carbon to another, use less current per candle-power than those of the incandescent form, but the candle-power of an arc lamp is, of course, immensely greater than that of a glow lamp. The E.M.F. required for each arc lamp, if used singly, is about 50 volts, but five lamps may be placed in series on a 240-volt circuit. The current used is from 5 to 20 amperes, depending upon the size of the carbons and candle-power of the lamps.

A 2,000 candle-power arc lamp of ordinary or open type absorbs about 500 watts, so that in actual practice the engine driving the generator must indicate nearly 1 horse-power for every 2,000 candle-power arc lamp.

The carbons in an ordinary open type of arc lamp require to be replaced after burning for 12 to 18 hours, but in the closed form of arc lamp, such as the Jandus, where the arc is enclosed and is maintained in an atmosphere of carbon monoxide gas and nitrogen, the carbons will last for 120 to 200 hours, depending upon the size of the carbons and the amount of current sent through them. From 5·5 to 7 amperes at 100 to 120 volts are required with this form of lamp.

To return once more to the generator. In the earlier stages of electric development, the two-pole horse-shoe form of dynamo was considered the best for machines giving an output up to 200 kilowatts, but when larger machines than this were called for, it was found that less material was required in dynamos having a considerable number of poles, and multipolar machines have now generally superseded bipolar machines even in the smaller sizes.

Fig. 78 gives the end view of a 750-kilowatt multipolar dynamo without its end bearing. In a multipolar dynamo, as shown, the yoke and magnets are made of high permeability cast steel—*i.e.*, steel which presents very small resistance to the lines of magnetic force. The poles being alternately north and south, the lines of magnetic force do not require to travel right across the armature as in a two-pole dynamo, but merely from one north pole through a portion of the soft iron core of the armature, and into the adjoining south pole.

In a direct-current generator giving current at a high voltage, it is necessary to have a large number of segments in the commutator in order to keep the difference of potential between each segment as small as possible, and thus avoid any danger of the current jumping across the insulation between the segments. If the commutator is of large diameter and the speed is high, the centrifugal force of the segments is considerable. If, on the other hand, the diameter of the commutator is small, there is a risk of the current jumping from one set of brushes to the next. Difficulties in connection with the commutators of direct-current, high-voltage generators are not uncommon, especially when such generators are coupled to steam turbines.

Direct current at a high E.M.F. can be obtained by working two or more dynamos in series—*i.e.*, the first dynamo supplies current to the second, the second to the third, and so on. Thus two dynamos, each giving 50 amperes at 3,000 volts, will, if coupled up in series, give 50 amperes at 6,000 volts, and three such machines 50 amperes at 9,000 volts. This system was first carried out on a practical scale by M. Thury, and is known as the Thury system.

The reason why high voltages are required is this—a current of 1,000 amperes at a pressure of 10 volts has the same energy as a current of 10 amperes at a pressure of 1,000 volts, but the former current requires a conductor having about $\frac{1}{4}$ of a square inch sectional area to carry it, while a current of 10 amperes

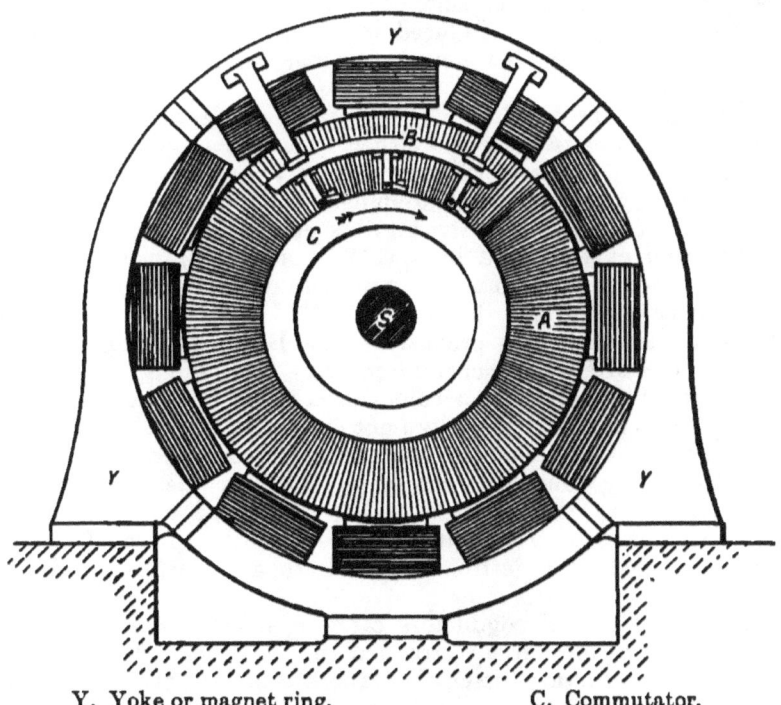

Y, Yoke or magnet ring.
B, Part of brush ring carrying three brushes.
C, Commutator.
S, Shaft.

Fig. 78.—Twelve-pole direct-current dynamo.

requires a conductor having only $\frac{1}{400}$ part of a square inch to carry it, and as copper is very expensive, it is necessary, when current has to be transmitted to a great distance, to transmit it at as high voltage and small amperage as possible.

An alternator is much more suitable for producing current at a high E.M.F. than a direct-current generator, as it requires no commutator. An alternating current can be used for lighting purposes, provided the alternations are sufficiently rapid—viz., from 40 to 50 per second; an alternating current also has the following valuable property:—If two conductors are placed side by side, and an alternating current is sent through one of them, it will induce a current (having alternations in the opposite direction) in the other conductor, and the E.M.F. of the secondary or induced current can be made either higher or lower than that of the primary current. Thus, if a wire of large diameter, carrying a current of a good many amperes, is placed alongside a smaller wire, and having a greater resistance, the current induced in the latter will be of higher voltage, but of smaller amperage, than in the primary wire, and *vice versâ*.

By the aid of a suitable transformer, an alternating current can therefore be transformed either to a higher or lower E.M.F. A continuous or direct current will only induce a current in a second wire at starting and stopping. In lighting a town having a number of outlying districts, it is customary to generate an alternating current at a high E.M.F., and to transmit at this high tension to a number of transformer stations, where the current is transformed down to the E.M.F. at which it will be used on the consumer's premises. The Board of Trade will not allow the E.M.F. of any current entering a private house or shop to exceed 500 volts. As a rule, the E.M.F. of the current used in a private house or shop does not exceed 220 or 240 volts.

If a current has to be transmitted to a distance of many miles, and the voltage decided upon is too high for the insulation of the alternator, the current can be generated at a moderate E.M.F. and then transformed up (by a step-up transformer), transmitted to its destination, and then transformed down again.

The insulation of conductors carrying current at a very high E.M.F. can be more easily arranged in a stationary transformer than in the generator itself, but there is a loss every time the current is transformed up or down.

Alternators, Single and Multi-phase.—In a large alternator, where, as we have seen, it is not necessary to have a commutator, it is customary to rotate the field magnets, and to have a fixed armature. A large number of poles, alternately north and south, are arranged on the rotating magnet, while the armature is outside it. Fig. 79 shows the arrangement diagrammatically with a uni-coil winding—*i.e.*, one slot or coil per pole piece. Plan 1 shows the direction of the current in one of the armature coils while a north pole is passing under the right-hand portion of the coil, while Plan 2 shows the reversal of the current while a south pole is passing under the left-hand portion of the coil.

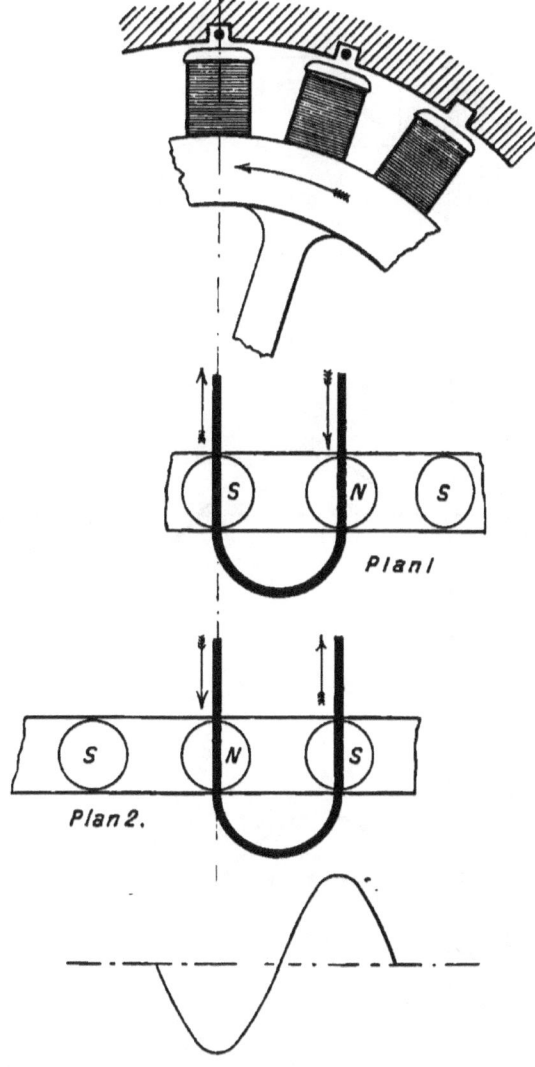

Fig. 79.—Alternator.

Two- and Three-phase Alternators.—An ordinary single-phase alternating current has one great disadvantage—viz., it will not start an electric motor. This difficulty in connection with the alternating current has, however, been overcome by employing three-phase alternators and motors. The meaning of

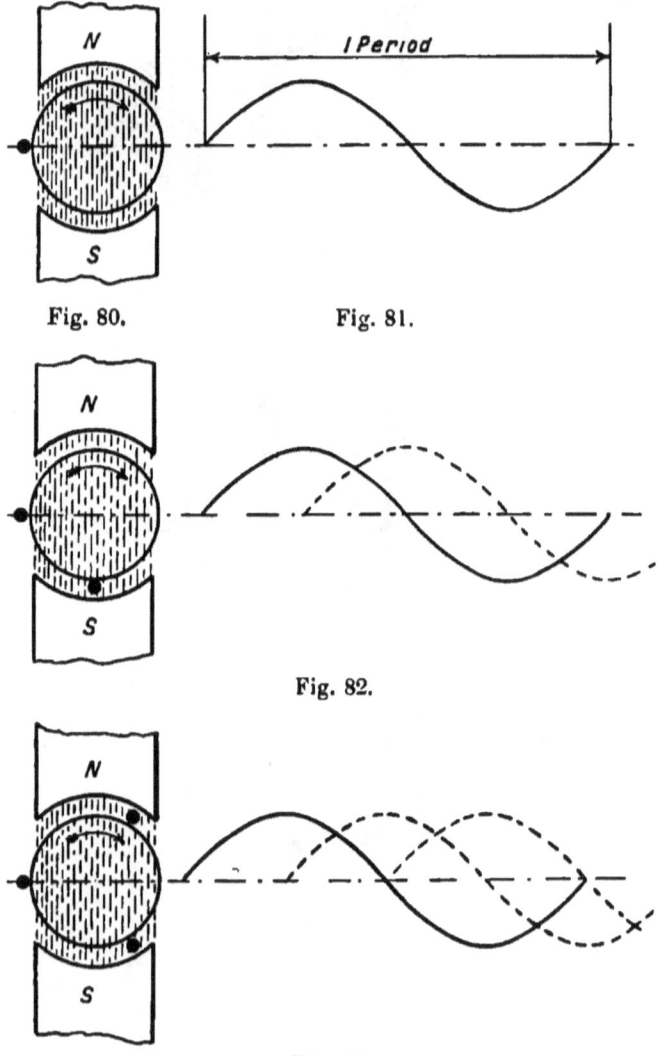

Fig. 80. Fig. 81.

Fig. 82.

Fig. 83.
Alternating current diagrams.

a three-phase alternating current can best be explained by the help of diagrams.

Fig. 80 represents diagrammatically a simple form of single-phase alternator having a north and south pole, and a single conductor on the drum, and Fig. 81 represents graphically the alternating current produced by one revolution of the drum carrying the

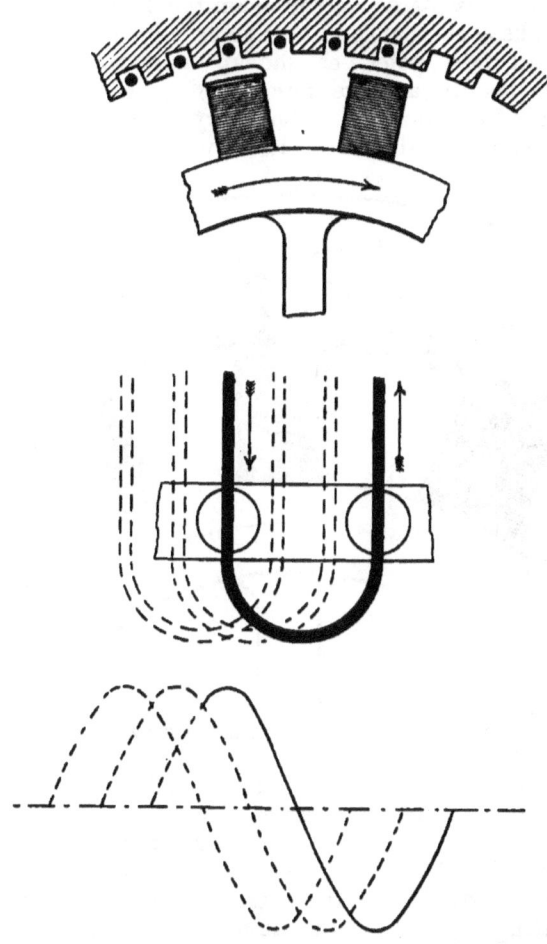

Fig. 84.—Alternate current diagram.

conductor; the highest point of the curve represents the greatest E.M.F. of the current, and the lowest point the zero. One revolution of the drum produces one complete period, so that, if the drum revolves 50 times in a second, the periodicity of the alternator (having one pair of poles) will be 50 complete periods per second. If a second conductor is placed on the drum the current produced will be as shown by Fig. 82, and will be a two-phase current. If the number of revolutions of the drum remains the same as before, the periodicity of the two-phase current will also remain unchanged—viz., 50 per second. If a third conductor is placed on the drum the resulting current will be a three-phase current, and will be as shown by Fig. 83.

Instead of two poles, as shown, there are, as previously mentioned, usually a large number of poles; these poles are rotated while the armature is stationary. The arrangement of a uni-coil three-phase alternator is shown by Fig. 84. In actual

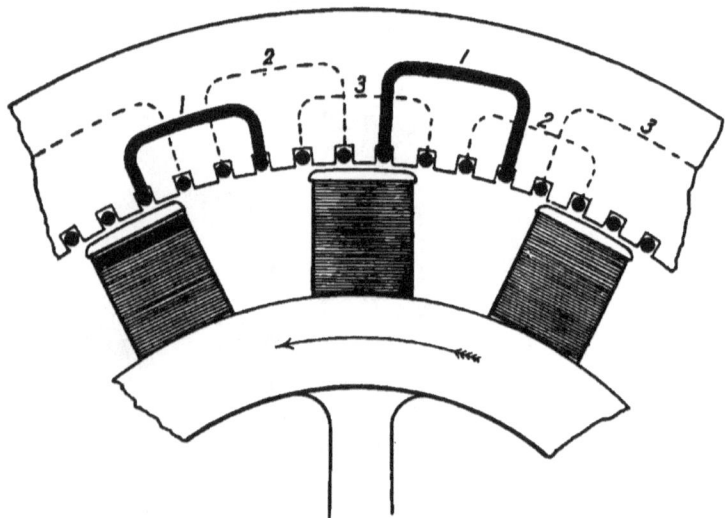

Fig. 85.—Alternating current generator.

practice, however, the two legs of the U-coil are seldom separated so widely as to come one over a North pole and one over a South pole. The winding of a three-phase alternator varies greatly, but Fig. 85 is a good example of such winding.

Periodicity of Alternating Currents.—The periodicity of an alternating current, if the number of poles and number of revolutions of the alternator are known, may be found thus:—

$$P = N \times R;$$

where P = periodicity per second.
N = number of *pairs* of poles.
R = revolutions per second.

If it is desired to find the number of revolutions *per minute* an engine or turbine must make to give a certain periodicity per second, the formula is—

$$R = \frac{P \times 60}{N};$$

where R = revolutions per minute.
P = periodicity per second.
N = number of pairs of poles.

Example.—Suppose an alternator, which has to be driven by a turbine, has 4 poles (2 pairs), and the periodicity required is 50 per second, at what speed must the turbine run? The calculation is—

$$\frac{50 \times 60}{2} = 1,500 \text{ revs. per minute.}$$

Or, suppose an alternator, requiring to be driven by a high-speed engine, has 20 poles (10 pairs), and the periodicity required is 50 per second, at what speed must the engine run?

$$\frac{50 \times 60}{10} = 300 \text{ revs. per minute.}$$

Before leaving the subject of generators, it may be well to explain the difference between series-wound, shunt-wound, and compound-wound dynamos. A series-wound dynamo is one in which the whole of the current generated in the armature is passed round the magnets to excite them. Such a dynamo is seldom met with, as, if extra resistance is put into the main circuit through which the current is flowing, the current will be reduced, and the exciting current will also be reduced; the same thing happens if the speed of the dynamo falls off. In both these cases the exciting current is weakened just when it should really be strengthened. Series-wound *motors* are, however, used, as a strong field is required at starting.

A shunt-wound dynamo is one in which the current is split up when it leaves the collecting brushes; a portion of the current is sent round the magnets, and the remainder flows into the main circuit. If extra resistance is put into the main circuit, the current finds an easier path round by the field magnets, and so strengthens the field. The amount of current flowing round the field magnets is usually regulated by means of a resistance and switch; thus the strength of the field can be increased or decreased at will.

A compound-wound dynamo is one in which a few turns of series winding are placed over the shunt winding, and arranged to act in opposition to it. Thus, if the speed of the generator should fall and the main current be reduced, the opposing power of the series winding will also be reduced, and the shunt winding, freed from this opposition, will more strongly excite the magnets.

A separately-excited dynamo is one in which the exciting current is supplied from another dynamo, or from some independent source. All alternators require to be separately excited.

Rotary Converter and Motor Generator.—If alternating current is employed for transmitting energy to a distance, and it is required to convert the alternating current into continuous or direct current, this can be done in two ways. The first is by means of a rotary converter. This consists of a motor, the armature of which is driven by the alternating current; over the winding required for this alternating current is placed winding suitable for producing continuous current. When the armature is rotated, the latter winding produces direct current in the same way as in a direct-current generator.

A motor generator consists of an alternating-current motor coupled to and driving a continuous-current generator. A motor generator is practically the same thing as a rotary transformer, only the alternating-current motor and the direct-current generator are two distinct machines coupled together, instead of being combined, as in the case of the rotary transformer. It is found that better regulation of the E.M.F. can be obtained by the use of two separate machines.

Primary Batteries and Accumulators.—A small current at a low E.M.F. can be obtained from primary batteries or cells, but these, although useful for electric bells and telephone work, hardly concern the mechanical engineer, as the production of current on a large scale by batteries is out of the question on account of the cost.

The simplest form of cell, known as a Daniell cell, consists of a plate of zinc and one of copper immersed in dilute sulphuric acid; the acid eats away the zinc, and, if the zinc and copper plates are connected by a wire, a current flows from one plate to the other through the acid and through the wire. The E.M.F. of such a cell is 1·05 volts. In the Leclanché cell, largely used for electric bells, there is a rod of zinc, also a porous pot containing a block of carbon, small pieces of crushed carbon, and black oxide of manganese; these replace the copper plate of the Daniell cell. The zinc and porous pot are immersed in a solution of sal-ammoniac. The E.M.F. of this cell is 1·4 volts.

Accumulators.—Accumulators are used for storing electricity, if such an expression may be used, for the action is really a chemical one. An accumulator consists of two or more plates, positive and negative, pierced by a large number of holes. The holes in the negative plate are usually filled with pellets of lead oxide, while the holes in the positive plates are filled with pellets of peroxide of lead; these plates are immersed in dilute sulphuric acid; a current is passed from one plate to the other through the acid, when a certain chemical change takes place. After the current has passed for a sufficient length of time, the accumulator is said to be charged. That is to say, if the plates are connected together by a wire, a current will flow in the direction opposite to that in which the current entered the accumulator. Each cell is charged until the E.M.F. reaches about 2·5 or 2·6 volts; when the charging wires are disconnected, the E.M.F. falls to about 2·1 volts, and as the accumulator discharges its current the E.M.F. gradually falls to 1·85 volts, when the accumulator is practically discharged. Any further discharge is injurious to the accumulator.

The capacity of an accumulator in amperes depends upon the surface of the plate or plates. In practice there are several plates in each cell; all the positive plates are connected to one bar, and all the negative plates to another bar. To obtain a higher E.M.F. than 2·1 volts, a number of accumulators may be placed in series; in this case the positive terminal of one accumulator is connected to the negative terminal of the next, and so on. Thus, if two accumulators are placed in series, the E.M.F., when charged, is 4·2 volts*; if 10 accumulators are placed in series, the E.M.F. is 21 volts; and if 50 are in series, the E.M.F. is 105 volts. If it is essential to maintain the total E.M.F. constant while the accumulators are being discharged, a few additional cells must be added to the series to compensate for the falling off of E.M.F.

In order to charge accumulators, direct current must of course be used, and the E.M.F. of the current must be a little higher than that of the whole of the cells being charged in series. Accumulators may be coupled up in parallel; this means that the whole of the positive plates are connected together, and the whole of the negative plates together. If coupled up in this way, the total E.M.F. is only that of a single cell, but the total current in amperes is increased in proportion to the number of cells employed.

The capacity of an accumulator is known by ampere-hours; thus a 24 ampere-hour accumulator will give 1 ampere for twenty-four hours, or 2 amperes for twelve hours, or 4 amperes for six hours. The rate of charging and discharging

* The 4-volt accumulators used for motor-car ignition are really two 2·1 cells coupled in series in one case.

is stated by the makers, and should not be greatly exceeded; if it is, gas is formed too quickly, and tends to dislodge the pellets.

The acid used consists of 8 parts of sulphuric acid to 5 parts of water, and has a density of 1·2. As the accumulator discharges the acid becomes weaker, so that by placing a hydrometer in the acid and noting how far it sinks, the amount of charge in the accumulator may be known approximately.

Accumulators are very useful for country house lighting, as they can be charged in the day time, and the current used during the evening and night time. They are also useful in small central stations as a reserve, and to maintain a supply of current during the night, should the demand not be sufficient to warrant the running of an engine and dynamo. Accumulators are also used sometimes as a steadier for the current; in this way,—the engine and dynamo supply the main demand for current, any excess current going through the accumulators; should the demand be suddenly greater than the engine and dynamo are capable of meeting and the E.M.F. falls, then the accumulators automatically discharge their current, and so assist the engine and dynamo. When the demand falls off and the E.M.F. rises, some of the current from the dynamo flows round by the accumulators and recharges them.

The efficiency of an accumulator is from 70 to 80 per cent.—that is to say, it will return 70 or 80 per cent. of the current put into it.

The disadvantages of accumulators are—(1) Their first cost. This is somewhat high, as for an E.M.F. of 120 volts about 70 accumulators are required, as one can only reckon upon the E.M.F. of the cells when nearly discharged—viz., 1·85 volts; (2) the cost of upkeep, which is rather high; and (3) their great weight. Objections 2 and 3 prevent accumulators coming into general use for traction purposes.

The instruments used for measuring the electric current are the voltmeter, for measuring the E.M.F.; the ammeter (or ampere meter), for measuring the amount of current flowing; and the wattmeter; the latter measures the volts multiplied by the amperes. An automatic cut-out is an instrument for automatically breaking the circuit should an excessive current pass.

A summary of the electrical units, with which it is desirable that the mechanical engineer should be familiar, is as follows:—

ELECTRICAL CHAPTER.

Ampere = unit of current.
Volt = unit of pressure or electromotive force (E.M.F.).
Ohm = unit of resistance.
1 ampere × 1 volt = 1 watt.
746 watts = 1 electrical horse-power.
1,000 watts = 1 kilowatt.
1 megohm = one million ohms.

$$\frac{\text{Volts}}{\text{Ohms}} = \text{amperes.} \qquad \frac{\text{Volts}}{\text{Amperes}} = \text{ohms.}$$

Amperes × ohms = volts.

The following table giving the resistances in ohms per 1,000 yards of copper wires of different diameters, and the amount of current in amperes carried, may possibly be useful. The current given is based upon 4,000 amperes per square inch of sectional area. The current carried by a cable having several strands can also be ascertained by the table—thus, a $\frac{3}{20}$ cable will carry a current three times greater than a single No. 20 wire, or 12 amperes. A $\frac{7}{20}$ cable will carry 28 amperes, &c.:—

TABLE XVIII.

S.W.G.	Diameter.		Sectional Area.		Resistance in 1,000 Yards.	Working Current at 4,000 Amperes per Sq. In.
	Inch.	Mm.	Inch.	Mm.	Ohms.	Amperes.
22	·028	·711	·0006	·397	38·46	2·4
20	·036	·914	·0010	·657	23·26	4·0
18	·048	1·22	·0018	1·167	13·10	7·2
16	·064	1·62	·0032	2·075	7·36	12·8
15	·072	1·83	·0040	2·627	5·81	16·0
14	·080	2·03	·0050	3·243	4·71	20·0
13	·092	2·34	·0066	4·287	3·57	26·6
12	·104	2·64	·0085	5·48	2·78	34·0
11	·116	2·95	·0106	6·818	2·24	42·4
10	·128	3·25	·0129	8·302	1·84	51·5
9	·144	3·66	·0163	10·507	1·45	65·2
8	·160	4·06	·0201	12·972	1·18	80·0
7	·176	4·47	·0243	15·695	·99	97·2
6	·192	4·88	·0289	18·679	·83	115·6

CHAPTER XII.

HYDRAULIC MACHINERY.

The words "hydraulic machinery" cover a somewhat wide field: we will first consider hydraulic machines, such as presses, riveters, and lifts, and then pass on to water-wheels, turbines, and pumps.

Power can be transmitted to considerable distances by means of water under pressure in a safe, clean, and fairly economical manner, and such transmission is usually employed in cases where large powers are required to be exerted during a short space of time and intermittently, as in hydraulic riveters, hydraulic presses for flanging boiler plates, &c. Hydraulic power is also largely used for cranes and lifts where the power, although not necessarily great, is required intermittently.

The principles governing the construction of hydraulic machinery in which water under high pressure is employed are fairly simple, none of the problems arising in connection with the use of steam, such as initial condensation, re-evaporation, &c., are met with, and if the amount of power required is known it is not a difficult matter to design a machine to give it.

The hydraulic press fitted with a small hand pump, illustrated by Fig. 86, shows the principle upon which most hydraulic presses, riveters, and lifts work. The action is as follows:—The ram of the press is of large diameter while the pump plunger is of very small diameter. The small plunger on being raised draws in water through the valve at the bottom of the pump; on being pressed down the plunger forces water through the valve at the side of the pump into the main cylinder which contains the ram of the press. Now, if the area of the ram is fifty times greater than that of the small plunger, every 1-lb. pressure exerted on the plunger will give a pressure of 50 lbs. on the ram: if a pressure of 100 lbs. is exerted on the small plunger a pressure of 5,000 lbs. will be given to the large ram.

Of course the large ram moves very slowly if the water is forced into the cylinder by a very small plunger, and in practice such a hand pump would not be used, but even with pumps driven by steam power the motion of the ram would be too slow if the water were pumped directly to the cylinder. This difficulty is overcome by pumping the water into an accumulator, storing it there under pressure, and drawing from the accumulator as required.

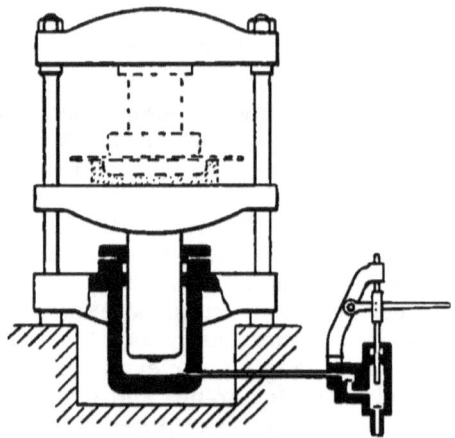

Fig. 86.—Hydraulic press.

A well-designed hydraulic accumulator having a ram 8 inches in diameter and a stroke of 14 feet is shown by Fig. 89. Drawings of accumulators are not frequently given in text

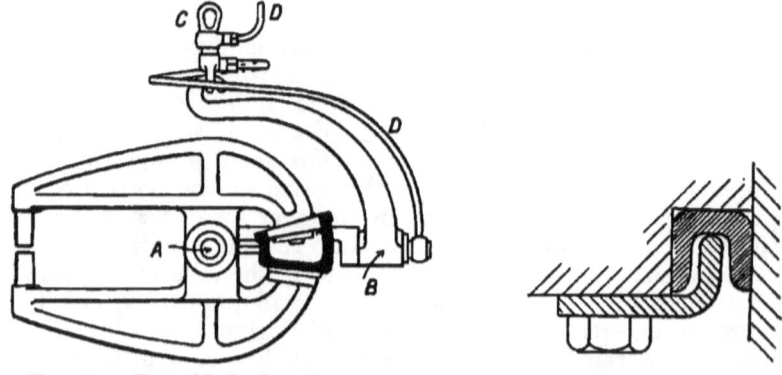

Fig. 87.—Portable hydraulic riveter. Fig. 88.

books or technical journals, and the reader would do well to examine the illustration carefully, comparing it with the small drawing of an accumulator which is taken from a work of a popular character. The accumulator illustrated on the larger scale shows the simplicity and directness characteristic of all the

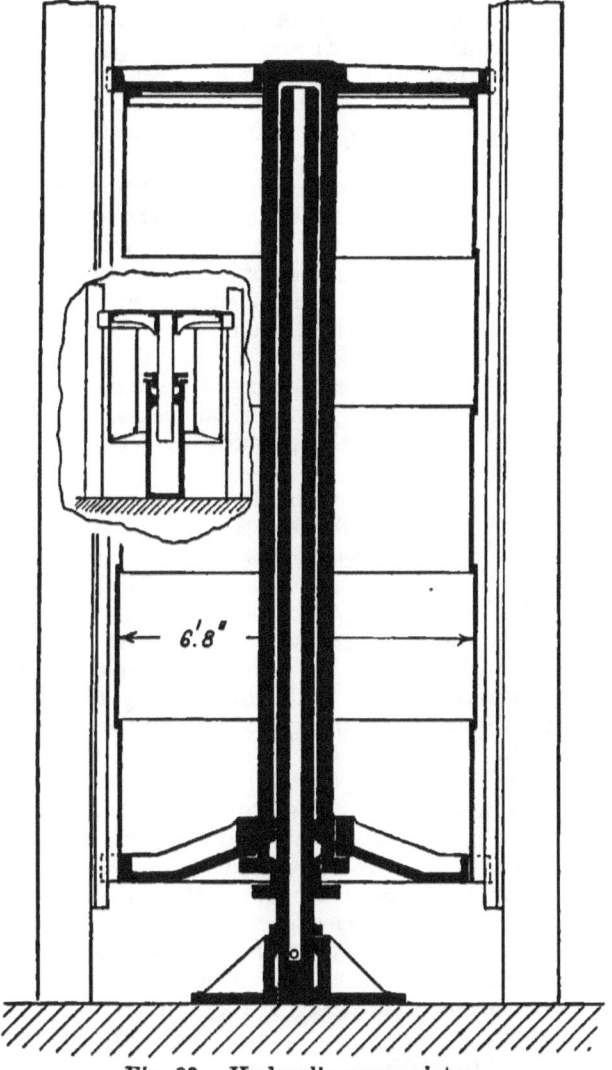

Fig. 89.—Hydraulic accumulator.

designs of the late Mr. Tweddell (and of Messrs. Fielding & Platt). It will be seen how, by inverting the cylinder, the gland has been made accessible for packing, how the weight in the casing is carried by the cylinder without the necessity for supporting it from a cross beam at the top, and how the cylinder walls form the inside of the casing, enabling a single outer casing to take the place of the inner and outer casings necessitated by a fixed cylinder and moving ram. In very large accumulators a supporting top cross girder is sometimes used, but the moving cylinder is, of course, retained. To proceed with the description:—The accumulator consists of a cylinder which is free to move up and down on a ram; the cylinder carries a large wrought-iron casing which is filled with scrap iron, slag, or any material sufficiently heavy to give the required weight and pressure per square inch on the water. When water under pressure is pumped into the accumulator the cylinder carrying the casing rises, and the water in the accumulator remains under pressure until it is required. When water is taken from the accumulator the cylinder and casing descend for a short distance, and a further supply of water is pumped in, and, as the capacity of the accumulator is much greater than that of the cylinder of any one hydraulic machine, there is always available a supply of water under pressure.

In practice steam-driven pumps are usually employed to force the water into the accumulator, and there is a simple device by means of which the pumps are stopped when the accumulator has reached the top of its stroke, and automatically started when the accumulator has fallen a certain distance.

The pressures used in hydraulic engineering vary from 750 lbs. per square inch to 2,240 lbs. per square inch. The pressure used by Mr. Tweddell, and still employed by Messrs. Fielding & Platt, in connection with hydraulic riveters, is 1,500 lbs. per square inch.

The credit for the introduction of hydraulic riveting machinery is due to the late Mr. R. H. Tweddell, and to him alone. This gentleman had considerable difficulty in getting any firm of engineers to take up and manufacture his plant. Shipbuilders, boilermakers, and others considered that it was impossible to make a riveter sufficiently portable to be of much use, and they doubted whether the cost of closing rivets by hydraulic power would be as low as was the case with hand-riveting. However, after Messrs. Fielding & Platt undertook the manufacture of the riveters, and their advantages became known, hydraulic riveting soon became general, and, at the present time, probably no boiler works or shipbuilding yard is without a hydraulic riveting plant. In addition to doing the work more quickly and cheaply than by hand, the work done is sounder and better. The comparison of a section cut through a riveted joint, the rivets of which have been hydraulically closed, with a section cut through a pair of plates and through

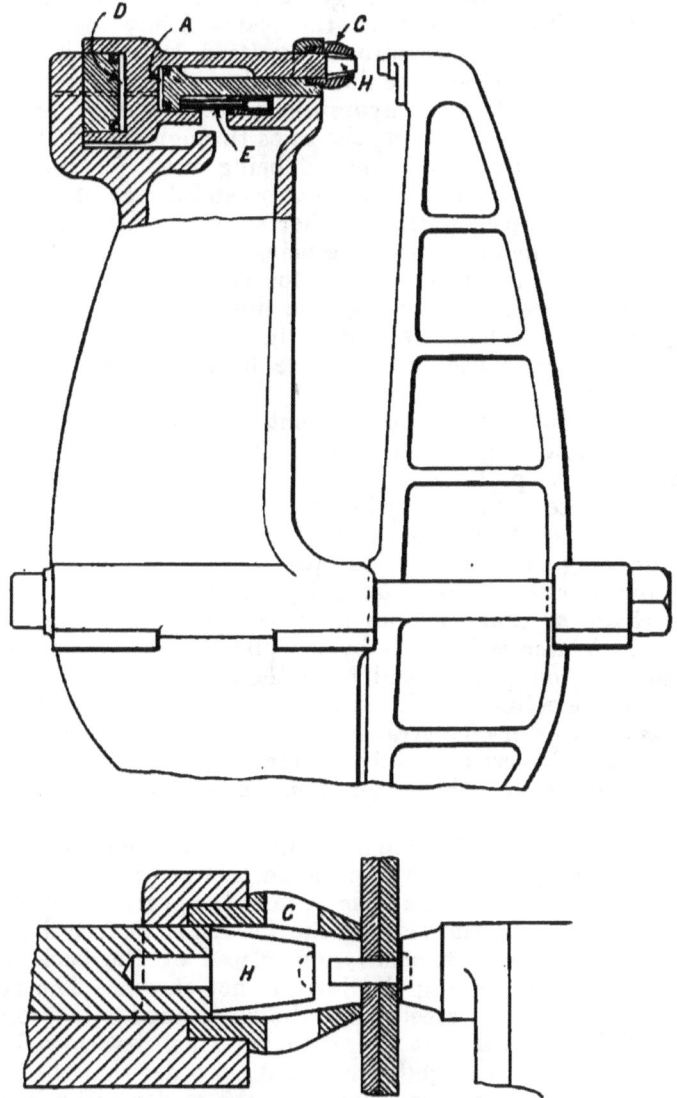

Fig. 90.—Hydraulic riveter.

rivets which have been closed by hand, shows the difference at once.

Fig. 90 shows a large fixed riveter and plate closer. Water is first admitted to the cylinder A, and moves forward the plate closing cup C (shown to a larger scale at the bottom of the illustration); this cup presses the plates firmly together, and holds them so while pressure is admitted to cylinder D; the effect is to move forward the die H, and to close the rivet.

In the earlier days of hydraulic riveters water under full pressure was used for moving the rams through the whole of the outward stroke, the small ram E being used for the return stroke only. This was a somewhat wasteful way of using the water, and modern riveters are either supplied with water under two pressures, the lower pressure being used to bring the die up to its work, and the full pressure for closing the rivet; or the small ram is used for bringing the die up to its work, water being permitted to flow into the cylinder from a tank 20 to 30 feet overhead, and the full pressure being used for doing the actual work.

In cases where it is inconvenient to bring the work, such as a long lattice girder, to the riveter, portable riveters are employed. A portable riveter is shown by Fig. 87. It will be noticed that the cylinder is placed at the end of the riveter farthest from the dies; this arrangement admits of the dies being used in confined spaces. The riveter can swivel at A and B, also at the hook C by which it is hung from the crane chain, so that the dies may be turned into any desired position. Water is brought by the pipe D; it is carried to the central gudgeon pin A, and from thence, at the back of the riveter to the cylinder. The pipe leading from the crane to D consists of a copper pipe arranged spirally round the chain; when the riveter is lowered the spiral extends; when it is raised the spiral closes; so that the water supply is not affected by raising or lowering the riveter.

The joint between the moving piston or ram, and the cylinder walls is made by a leather U-ring, an enlarged view of which is given in Fig. 88; with this form of joint the greater the pressure of water, the more the side of the U-leather is forced against the cylinder walls. A small ram placed at the back of the riveter, used for opening the jaws, is not shown. The riveter arms are made of cast steel; the cylinder is lined with bronze.

It will be noticed that the cylinder must be curved at a radius struck from the gudgeon pin A, in order that the riveter arms may move about this centre. If the reader will ask any of his engineering friends how such a cylinder is bored, two out of three will probably say that it can only be done by special machinery, and it is so stated in a well-known text-book. This, however, is not correct; the curved cylinder at the end of a riveter arm can be truly bored in an ordinary lathe, and, although this book does not deal with mechanical processes an exception will be made in the present case. The method of boring which the author saw in use some years ago is as follows:—The boring tool is attached to the lathe face plate, an upright pillar of the size of the gudgeon pin A is bolted to the saddle of the lathe, the riveter arm is carried on this pillar horizontally, with its cylinder towards the tool. The riveter arm is free to turn on the pillar, and the latter is placed at such a distance from the centre of the lathe that the tool as it revolves will take a cut out of the cylinder if the latter is pressed forward. A piece of hard wood, about a couple of feet long, is placed at the back of the cylinder, and carried to the back centre of the lathe. The lathe is set to work, the tool revolves, and the turner feeds the cylinder forward by means of the hand-wheel on the back centre; the arm carrying the cylinder being pivoted on the pillar referred to, the cylinder is bored to the right radius.

Hydraulic Flanging Press.—When a boiler plate to be flanged is not of too great a diameter it is done in a press similar to that shown by Fig. 86, but a powerful press may have two or three rams and cylinders in place of the single ram shown. The plate is made red hot and is placed between dies, as shown by dotted lines; the plate is flanged at one operation by the rising of the ram.

In cases where the plate to be flanged is of irregular shape and cannot well be done in the manner described, the flanging is done a piece at a time by means of a press, as shown by Fig. 91. One ram holds the plate, while the second ram turns it over; sometimes there is a third ram placed horizontally so that it may square up the flange which has been turned over. In working with these presses it is necessary to see that the plate is not allowed to get too cold while the operations are being carried on, as it has been found that steel worked at a blue heat loses a large proportion of its strength.

Fig. 92 shows a hydraulic punching machine; this hardly needs a description as its action is the same as that of the fixed riveter, but without the plate-closing arrangement. Hydraulic shearing machines working upon the same principle are also used in boiler works and ship yards.

Hydraulic Lifts and Cranes.—There are two methods of raising the cage of a hydraulic lift. The first is to place it directly on a ram, such as is shown in the illustration of the hydraulic press. This method is simple and safe, but necessitates a ram and cylinder of great length; it also necessitates a very deep pit to receive the cylinder. The weight, too, of such a ram is somewhat great and a considerable amount of power is expended in raising it; the weight of the ram can, of course, be balanced by counter weights and chains, but these introduce an element of risk and are very undesirable. A hydraulic balancer somewhat similar to an accumulator is sometimes used, into which some of the water is returned on the down stroke of the ram.

The second method of raising the cage of a hydraulic lift and one which is usually employed in hydraulic cranes, is by means of a jigger or multiplier as shown by Fig. 93. By the use of a jigger a weight can be lifted through a considerable distance with a comparatively short ram and cylinder. The chain is passed over multiplying sheaves, and every sheave on the ram multiplies

Fig. 91.—Hydraulic flanging machine.

Fig. 92.—Hydraulic punching machine.

the motion by two; the reason for this will be seen by glancing at the illustration, as for every foot by which the ram rises there must be a foot of chain on each side of the sheave. Thus if there are three sheaves and the ram rises a foot, a motion of 6 feet will be given to a weight at the end of a chain. A certain amount of power is lost by the friction of the chain and pulleys, but the convenience of this method of multiplying the lift outweighs the disadvantage. Rams arranged in this way—viz., with the chain passing over sheaves—are used for slewing or turning cranes, as well as for raising the weight.

Hydraulic Jack.—A hydraulic jack, as shown by Fig. 94, is very useful for raising heavy weights when these cannot be dealt with by a crane. The principle upon which the jack works is the same as that of the press shown by Fig. 86, but the jack is self-contained; the pump is placed inside the upper portion of the casing; this portion also holds the water which the pump plunger forces down into the space A, and causes the upper portion of the jack to rise. The weight to be raised is placed either on the top of the jack or on the projecting piece B. It was by the aid of such jacks that the steamer "Great Eastern" was successfully launched after having resisted all previous attempts

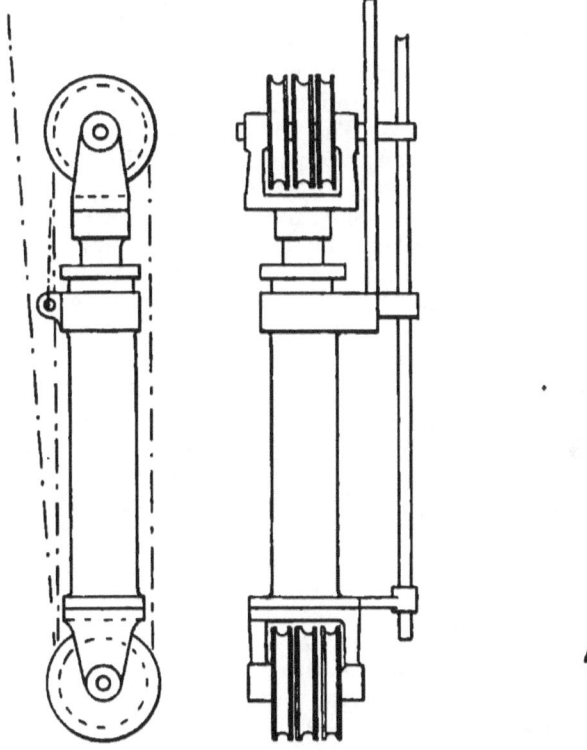

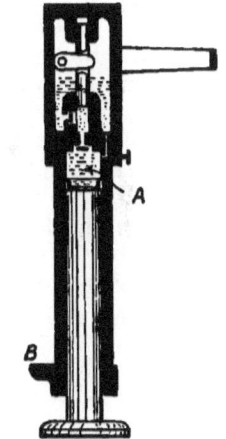

Fig. 93.—Hydraulic jigger. Fig. 94.—Hydraulic jack.

to move her. The Menai tubular bridge, too, was raised into position, a few inches at a time, by means of hydraulic jacks.

Hydraulic Tests.—Parts of machinery, such as steam cylinders, valves, boilers, hydraulic cylinders, &c., which are required to withstand pressure, are usually tested hydraulically, a hand pump such as is shown in Fig. 86 being used. In such tests it is necessary that the highest part of the piece under test should be provided with an air cock, and the whole of the air be expelled before the full pressure is applied. If this is not done, and the piece under test fails to bear the requisite pressure, the compressed air may cause broken pieces to fly in every direction, possibly with serious results. Water, unlike air, is practically incompressible, and in the event of the failure of the piece under hydraulic test, a loud noise is heard, the water escapes, but no damage is done, *provided no air is present.*

Water Wheels and Turbines.—In the days before steam power was available, mills, in which mechanical power was required, were built on the banks of streams and rivers, in places where a fall of water could be obtained. If the natural fall was not sufficiently great, it was sometimes increased by placing a dam across the stream, and thus raising the level of the water on one side of the dam. The water at the higher level then had potential energy, or energy of position, and this energy was converted into mechanical work by means of water wheels.

Water wheels are seldom constructed now, as the capital which has to be expended in constructing the necessary masonry work and wheel is usually greater than that needed to purchase a small engine and boiler capable of giving as much power as the cumbrous wheel, and the interest on the capital saved may go some way towards providing coal for the steam plant.

In cases where an adequate supply of water with a good fall is available, and considerable power is required, water turbines are employed in preference to water wheels; they are much less cumbrous, require less masonry, are more efficient, and can be governed more accurately than the latter. The leading features of water wheels will, therefore, only be described briefly before passing on to turbines.

Water wheels are of three kinds—viz., Undershot, Overshot, and Breast wheels. Undershot wheels are used in cases where the fall is low, say, 1 to 2 feet, and where a fair amount of water is running to waste. As its name implies, the undershot wheel is one in which the water strikes the vanes of the wheel at a point below the axis, and causes the wheel to rotate. The efficiency of an ordinary undershot wheel is very low—viz., from 25 to 30 per cent.—but if the buckets are curved, as in the "Poncelet" wheel, a much higher efficiency is obtained. The Poncelet undershot wheel partakes somewhat of the nature of a turbine.

Overshot wheels are employed in cases where the fall is sufficiently great, 15 feet or more, to allow of the water passing over the top of the wheel and descending on the far side. In this type of wheel the vanes are shaped like buckets, so that the weight of the water, as well as the energy with which it enters the buckets, is utilised for turning the wheel. An efficiency of from 60 to 75 per cent. can be obtained with this form of wheel.

Breast wheels are used in cases where the fall, 5 to 15 feet, is hardly sufficient to enable an overshot wheel to be used, and where the velocity of the water is not great. In the breast wheel the water is delivered just above the centre line or axis of the wheel, and its weight only is used for turning the wheel. The efficiency of an ordinary breast wheel is from 30 to 50 per cent.

If the amount of water flowing per minute, the distance through which it falls, and the efficiency of the water wheel are known, it is an easy matter to calculate the amount of power which can be obtained. An example is given in connection with turbines later.

Turbines.—The turbines chiefly used in Europe and in the United States of America are of the reaction type, and the meaning of the words reaction turbine is as follows :—We have read that, according to Newton's first law of motion, "every body continues in a state of rest or of uniform motion in a straight line, except so far as it is compelled by force to change that state." Now, if a body of water is moving in a certain direction, and is compelled to change that direction, force is required to effect the change; in the case of a reaction turbine, the vanes of the moving wheel are arranged (see Fig. 95) so as to compel the water to change its direction, and the resistance offered by the water to this change of direction exerts a reactionary force on the vanes of the moving turbine wheel; it is this reactionary force which drives the wheel.

In the impulse form of turbine, the buckets of the wheel are arranged, not so much with a view to change the direction of the jet of water, as to oppose its progress and bring it to a state of rest. The energy which the water has acquired in falling from a height—viz., kinetic energy, or energy of motion, is given up to the buckets, and is utilised for turning the wheel.

Parallel Flow or Axial Turbine.—Fig. 95 shows diagrammatically a turbine of this type, in which the water flows in a direction parallel with the axis of the turbine. This form of turbine is generally known as the Jonval, from the name of the engineer who introduced it. The illustration is almost self-explanatory; the water descends through the fixed guide vanes, which give it a certain direction; the moving vanes prevent the water from continuing in its proper course and deflect it; the reactionary force imparted to the vanes causes the wheel to rotate. This form of turbine is usually governed by throttling the water as it leaves the suction tube A.

Radial Inward Flow Turbine.—Fig. 96 shows a turbine

of this type, in which the water flows into the turbine wheel in a direction at right angles to the axis of the wheel. This form of turbine was introduced by Lord Kelvin.

Radial Outward Flow Turbine.—Fig. 97 shows a turbine of this type, in which the water flows in a direction radial to the axis, but outwardly.

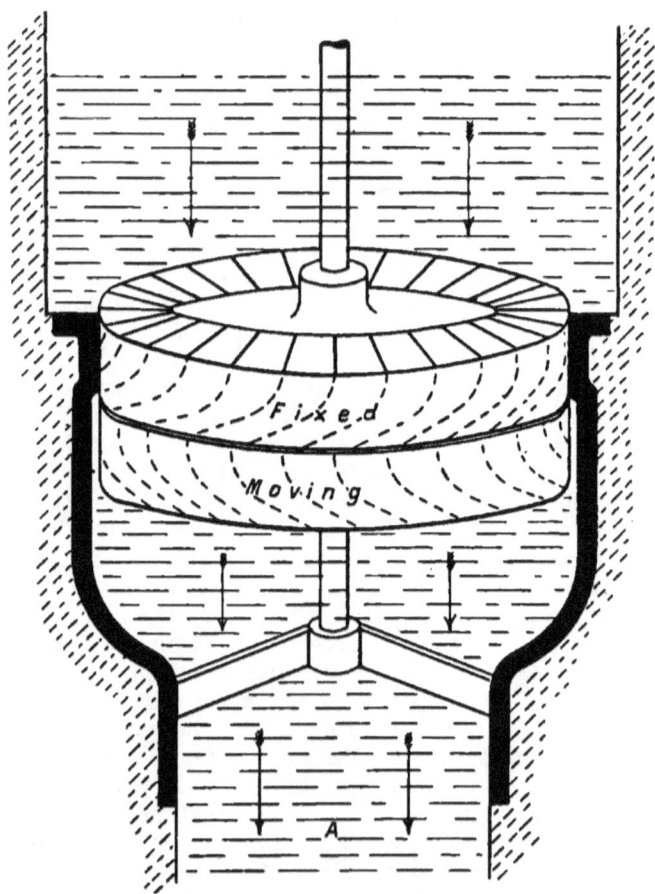

Fig. 95.—Parallel flow or axial turbine (Jonval type).

In addition to the three reaction turbines illustrated, there is the **Mixed flow turbine**, which is practically a combination of those shown by Figs. 95 and 96. In this turbine the moving blades, instead of ending as shown by Fig. 96, are prolonged and arranged so that they compel the inflowing water to change its course from a horizontal one to a vertical one.

Some of the advantages and disadvantages of each type of reaction turbine will now be considered.

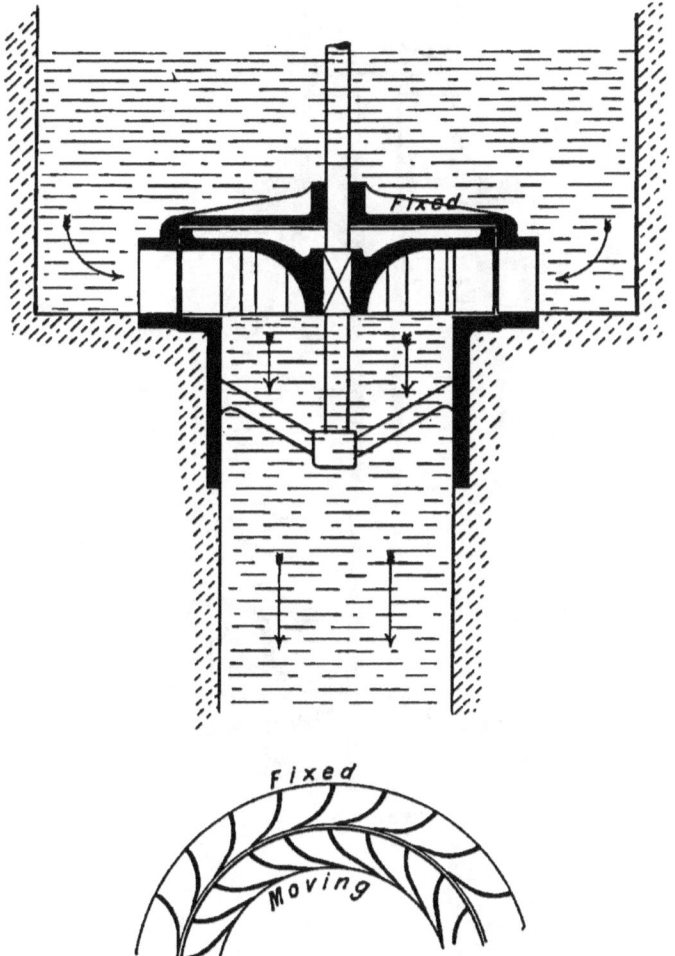

Fig. 96.—Radial inward flow turbine.

The outward flow turbine, shown by Fig. 97, was practically the earliest form of reaction turbine, and was introduced by a French engineer named Fourneyron. The chief disadvantage of this turbine is the difficulty of governing it with accuracy, and a further disadvantage is that it must be placed at the lowest point of the fall, near to the tail race. Turbines of this type

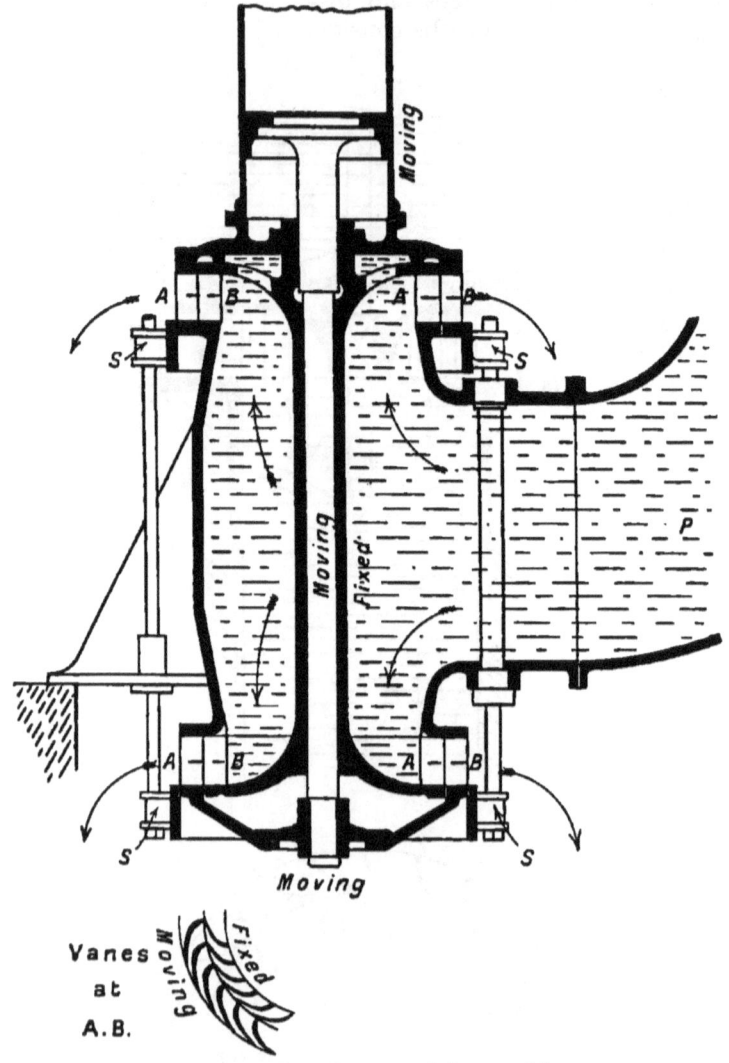

Fig. 97.—Radial outward flow turbine.

were employed in the first portion of what is probably the largest turbine installation in the world—viz., the Niagara Falls installation. The second instalment of turbines at this generating station were, however, of the inward flow type. This installation will be referred to again later. The efficiency of the outward flow reaction turbine is from 75 to 80 per cent.

The inward flow turbine has two advantages over that of the outward flow type, viz. :—(1) Its variation of speed can be controlled with greater accuracy, and (2) it may be used with a suction tube. The use of this tube enables the turbine to be placed (within certain limits given later) near the upper level of the water, the fall taking place in the suction tube. The efficiency of this turbine is from 75 to 80 per cent.

The parallel flow or Jonval turbine can also be governed with fair accuracy; it may be used with a suction tube, and gives a slightly higher efficiency than turbines of the radial flow type. The efficiency of a good Jonval turbine is from 80 to 85 per cent. The Jonval turbine, too, can be readily constructed for working in cases where the head of water varies. In such cases the turbine is provided with two or three rows of vanes, arranged concentrically, instead of a single row. When a good head of water is available the inner row is used, and the outer rows are shut off. When a small head only is available the outer row, or ring of vanes, is used.

The mixed flow turbine may also be used with a suction tube, will govern accurately, and has a still higher efficiency than a Jonval turbine. An efficiency as high as 86 per cent. is often obtained with a turbine of this type. The mixed flow turbine is much used in America; it may be of smaller dimensions for a given power than turbines of other types. The "Hercules" and "Little Giant" turbines are of this type.

Impulse Wheels.—These are frequently employed in mountainous countries, and are very suitable for use in cases where there is a very high fall of water, and where a moderate amount of power only is required. The most common form of impulse turbine is the Pelton wheel; this consists merely of a wheel, around the circumference of which are a number of buckets. In order to avoid shock when the water strikes the buckets, these are constructed in the form of the letter W, the upper part of the W points towards the jet, the jet strikes the central portion and glides down to the bottom of the bucket; the energy of the water is thus given up without shock. A Pelton wheel can be governed with great ease by throttling the jet. The largest wheel of this type which has been constructed gives about 700 B.H.P. The efficiency of a well-constructed Pelton wheel is from 80 to 85 per cent.

Before proceeding further, it may be well to show the student how to calculate the power which can be obtained with any given quantity of water falling through a given distance in a certain time. We have seen that 1 H.P. is equivalent to raising 33,000 lbs. 1 foot high in one minute. Now, if a body of water, weighing 33,000 lbs., falls through a distance of 1 foot, it is also the equivalent of 1 H.P., and if there were no loss in the turbine, 1 H.P. would be obtained from this quantity of water falling through the distance mentioned. Admitting, however, that there is a loss in the turbine, and assuming that the efficiency of the turbine is 75 per cent., then, if we multiply 1 H.P. by 75, and divide by 100, we get the actual horse-power —viz., ·75, or $\frac{3}{4}$ of 1 H.P.

The hydraulic horse-power given by a certain quantity of water falling through a certain distance may, therefore, be obtained by the following formula:—

$$\frac{C \times 62\cdot4 \times F}{33,000} = \text{H.P.};$$

where C = cubic feet of water per minute.
F = fall in feet.
H.P. = hydraulic horse-power.

If this hydraulic horse-power is multiplied by the efficiency of the turbine or water wheel, the actual or brake horse-power available for external work will be obtained.

For instance, if the hydraulic horse-power is 120 H.P., and the efficiency of the turbine is 80 per cent., the actual horse-power will be 120 × 80 ÷ 100, or 96 B.H.P. The reason for multiplying C by 62·4 is because a cubic foot of water weighs approximately 62·4 lbs. If the amount of water available is given in cubic feet of water per second, the figure must be multiplied by 60 to convert it into cubic feet per minute, or else the 33,000 lbs. must be divided by 60.

Example.—The amount of water delivered to each of the Niagara Falls turbines is 430 cubic feet per second, or 25,800 cubic feet per minute; the mean fall is 136 feet. What horse-power should be obtained? The calculation is

$$\frac{25,800 \times 62\cdot4 \times 136}{33,000} = 6,634 \text{ hydraulic horse-power.}$$

If we assume the efficiency of the turbines to be 75½ per cent., the actual horse-power available is 5,010. The turbines are supposed to give 5,000 B.H.P.

Suction Tube.—We have said that the use of a suction tube enables a turbine to be placed, within limits, near the upper level of the water. A suction tube, however, has no useful effect if it is over 30 feet in length, even if it is of small size;

with a suction tube of 5 feet diameter the limit of useful length is about 19 or 20 feet, with a tube 10 feet in diameter the tube should not be more than 10 or 12 feet long.

Governing.—The governing of a turbine with sufficient accuracy to make it suitable for electric lighting purposes is not altogether an easy task. If the regulation is effected by means of a sluice valve placed at the outlet of the suction tube, such a sluice valve requires to be opened or closed to a considerable extent before it materially affects the quantity of water passing. In some small turbines of the Jonval type the governing is effected by a device which rolls and unrolls a scroll of leather belting over the inlet vanes, thus cutting a certain number out of use when the turbine runs too fast. In the case of the Niagara Falls turbines the governing is effected by means of a circular sluice running right round the vanes, as shown by the letter S in Fig. 97. The sluice is shown fully open; by raising it the whole outflow of water can be stopped. In the Niagara turbines the vanes are divided into three equal portions which are successively closed as the sluice is raised. The vertical shaft of each of these turbines is provided with a flywheel 14 feet 6 inches in diameter, weighing 10 tons; the speed of the turbines is 250 revs. per minute. The rim of the flywheel is made of wrought iron, to enable it to withstand the stress due to centrifugal force. The reader may find it interesting to work out what the stress in the rim amounts to, by the rule given in Chapter VII.

As considerable power is required to move the sluices, the power is not taken from a centrifugal governor but from the turbine itself; a centrifugal governor is used to throw the regulating mechanism in and out of gear. At Niagara this arrangement was not found to act quickly enough to ensure very good governing, and at No. 2 power-house the sluices of the turbines are worked by separate hydraulic ram, the ram itself being controlled by a centrifugal governor. The sluices worked in this way can be completely opened or closed in ten seconds: the ordinary variation of speed does not exceed 1 per cent., and the momentary variation when the whole load is thrown off does not exceed 5 per cent.

In some turbines the vanes themselves are opened and closed in order to regulate the speed, but this method of governing introduces a certain amount of complication which is not altogether desirable in a turbine working under water.

Bearings.—The bearings of turbines of small size are often made of lignum vitæ and require no lubrication beyond that of the water; the friction, however, is greater than with metal bearings. With the latter it is usual to force in oil by means of a force pump. In many cases an over-head bearing is used which is not in contact with the water.

Niagara Falls Turbines.—These turbines, which were the largest in the world when constructed, have many interesting features. The turbines, ten in number, are designed to give 5,000 H.P. each, they are placed in a tunnel at the bottom of the fall, about 140 feet below the level of the room in which the electric generators are fixed. The turbines in No. 1 power-house have a double outflow, and are illustrated more or less diagrammatically by Fig. 97. They are constructed so that the pressure of the water in the turbine carries the great weight of the vertical shaft, flywheel, and rotating portion of the electric generator, amounting altogether to about 70 tons. The lower internal portion of the turbine casing is fixed and carries the downward pressure due to the column of water; the upper portion of the turbine casing has apertures so that the pressure of the water supports the upper moving wheel which carries the shaft, flywheel, and rotating portion of the dynamo. The mean fall is 136 feet, or, say, 130 feet, to the top of the turbine, and as every foot of a vertical column of water gives a pressure of ·434 lb. per square inch, the pressure under the moving wheel of the turbine is about 56 lbs. per square inch. The diameter of the moving upper portion of the turbine is 6 feet 3 inches, the area being 4,417 inches; multiplying this by 56 lbs. we find that the upward pressure of the water is over 100 tons. The shafts of these turbines consist of hollow steel tubes 38 inches in diameter and $\frac{3}{4}$ inch thick; these are lighter than solid shafts. The hollow shafts give place, however, to solid shafts 10 and 11 inches in diameter in the bearings.

The latter turbines installed in No. 2 power-house at Niagara are of the inward flow type.

Falls of Foyers Installation.—The largest turbine installation in this country is at the Falls of Foyers in Scotland. There are five impulse turbines with vertical shafts, each turbine gives 700 B.H.P. under a fall of 350 feet. The turbines are 9 feet in diameter and their speed is 140 revs. per minute.

Pumps for raising water are usually of the centrifugal or reciprocating types, but there is a third form of pump of which the best known example is the Pulsometer.

Centrifugal Pumps.—A centrifugal pump is shown by Fig. 98. This form of pump is suitable for dealing with large quantities of water when the height to which it has to be raised, or the head, is not very great. For heads up to 5 or 6 feet, a centrifugal pump is considered to be more efficient than any other type of pump; for heads up to 20 feet, it is considered to be as efficient as a good reciprocating pump; but when the head is over 20 feet a centrifugal pump is not so efficient as a good reciprocating pump, but it has other advantages, such as small first cost, small space occupied, ability to deal with muddy and gritty water, &c.

A centrifugal pump with a single impeller will raise water up to 60 or 70 feet, and by placing two such pumps in series—*i.e.*, one pumping into the other—water can be raised to a height of 140 feet or more. Pumps arranged in this way would not, however, work at the highest efficiency, and in cases where the head is greater than 70 feet the pump is usually constructed with two or more impellers in one casing, so that the passage from the

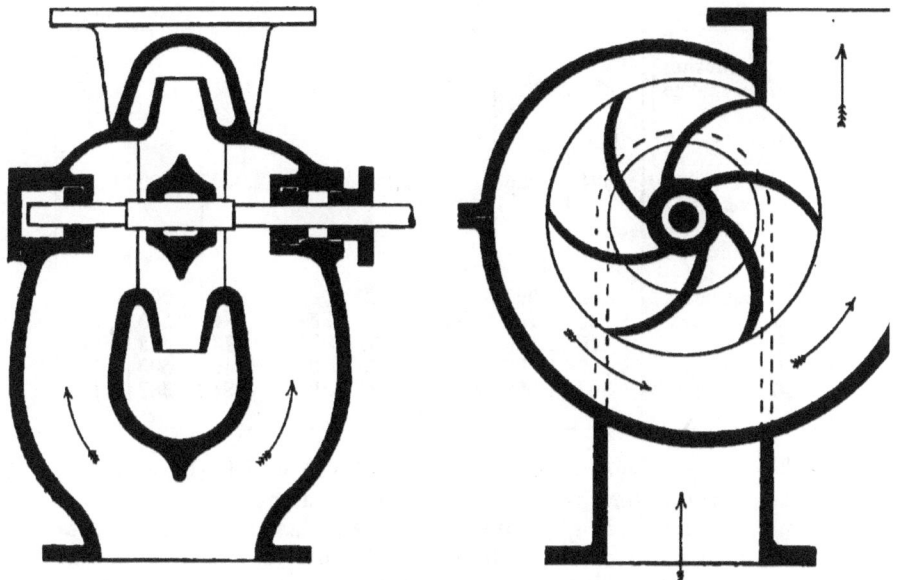

Fig. 98.—Centrifugal pump.

periphery of one impeller to the inlet of the next is as short and direct as possible.

High-speed multiple-stage centrifugal pumps, driven either by steam turbines or by electric motors, for delivering water against heads up to 500 and 600 feet are quite common on the Continent, and it is said that by coupling in series two four-stage pumps Messrs. Sultzer have been able to deliver water against a head of 1,700 feet. The pumps in question were driven by three-phase motors, the speed of motors and pumps being 1,040 revs. per minute.

A centrifugal pump, after having been charged, will continue to draw its water from a depth of 25 or 26 feet below the centre of the pump, provided the joints of the suction pipe are all tight, but such a great suction lift is not desirable. The suction lift should not be more than 12 or 15 feet if practicable.

The efficiency of a well-designed centrifugal pump working under suitable conditions is from 70 to 76 per cent.

The following table gives the approximate number of gallons delivered per minute by some centrifugal pumps of Messrs. Gwynnes' make, also the speeds at which the pumps must be driven to deliver water against various heads :— .

TABLE XIX.

Size of suction and discharge pipes, . .	3″	4″	6″	8″	10″	12″	18″	20″
Approximate number of gallons discharged per minute.	184	325	730	1,300	2,000	2,930	7,000	8,500
Revolutions per minute for—								
10-foot head,	900	900	676	522	450	450	268	268
20- ,, ,,	1,160	1.160	876	677	584	584	350	350
40- ,, ,,	1,546	1,546	1,161	897	773	773	463	463
50- ,, ,,	1,700	1,700	1,275	986	849	849	500	500
70- ,, ,,	1,963	1,963	1,473	1.139	981	981	587	587

The speeds given in the table are those at which the standard pumps will deliver the quantities of water mentioned at the best efficiencies, and will give the beginner an idea as to the speed at which such pumps must be driven, but centrifugal pumps can be specially designed for lower or higher speeds. Thus, if it were desired to couple a pump direct to a vertical engine, the speeds given in the table would be too high, and a pump having a larger disc to run at a lower speed would be supplied; if coupled to a steam turbine, the speed would be too low, and a pump having a smaller disc would be designed. There are, however, certain limits in each direction; if worked outside these, the efficiency of the pump falls off very rapidly.

Fig. 98 is almost self-explanatory. All that need be said is that the pump must be charged with water before starting; a foot valve is provided to prevent the charging water running away, and to avoid the necessity of recharging the pump every time it is stopped and started. The water enters and leaves the pump as shown by the arrows; the rotation of the impeller or disc with curved vanes imparts the necessary motion to the water.

The illustration shows a pump with shrouded disc—*i.e.*, one in which there is a rim of metal on each side of the vanes. This shrouding has the effect of reducing the friction between the water which is being rotated and the sides of the pump; it also strengthens the vanes. In unshrouded or open type disc pumps, the vanes are strengthened by central ribs or fins. In some shrouded pumps every alternate vane is not carried right down to the boss, but ends at the shrouding; this gives a freer entry to the disc.

Reciprocating Pumps.—All reciprocating pumps work on practically the same principle. The bucket or plunger draws water into the pump barrel through suction valves during one stroke, and forces it out through discharge valves during the next stroke. The chief differences between reciprocating pumps are the manner in which they are driven; the form of water piston—*i.e.*, whether bucket or plunger; and the design and arrangement of the valves. A horizontal reciprocating pump of the direct-acting type has been illustrated by Fig. 21, and the somewhat uneconomical method of driving it referred to. A vertical reciprocating pump, in which the valves are placed in the bucket, is illustrated by Fig. 64. Large pumps are usually driven by compound or triple-expansion engines, and extremely economical results are obtained. A consumption of coal of 1·4 lbs. per water horse-power is obtained in the best pumping plants.

The water piston or bucket is frequently replaced by a plunger or ram, which will displace the desired quantity of water per stroke without the necessity of touching the sides of the pump barrel. The plunger works through packing, and does not get cut or scored in the same way as a bucket. Mine pumps and other pumps which deal with gritty water almost invariably have plungers, but in mines reciprocating pumps are being largely superseded by centrifugal pumps.

Pumps must be placed within 25 feet of the level of the water to be pumped, and if the water is hot the pump should be placed below the level of the water; but even when this is done, there is usually some difficulty in pumping very hot water. The reason is this—We have seen that the point at which water turns into steam depends upon the pressure upon it, as well as upon the heat it contains. Now, if a pump attempts to lift water at 200° F. (and containing 1,142 B.T.U.), immediately the ram or bucket moves quickly away from the water, the latter tends to lag behind owing to inertia and to the friction in the pipes and valves, and thus a partial vacuum is created; the water then, instead of following the ram, turns into steam, and the pump has to deal with steam instead of water. It is for this reason that feed pumps are always arranged to force the feed water through an economiser, or an exhaust feed-water heater; and not to draw the water through, and then force it into the boiler.

The successful working of a pump depends very largely upon the valves employed. The earliest form of valve, called a clack valve, consisted of a leather flap; the leather formed the hinge, and the part of the leather which covered the opening was strengthened by a metal plate. Such valves were only suitable for low lifts and moderate speeds. A clack valve is sometimes made with a metal hinge. The pin is a very loose fit in its seat, so that the face of the valve may press tightly against the valve seat. Such valves are sometimes used in mine pumps where the water is very gritty.

A plain gun-metal disc valve, which rises off its seat and is brought back by the pressure of water above it, and by a spring, is used in hydraulic pressure pumps, but with this form of valve the pump must run at a low speed, and the valve must have a small lift; otherwise, a heavy blow is struck every time the valve closes. The duplex pump (Fig. 21) has, for the sake of illustration, been shown with two different types of metal valves, while three other types of valve have been shown on the drawing of the air pump (Fig. 64).

The head valves in Fig. 64 consist of a grid and guard, the valve itself being of rubber. The guard is saucer-shaped, and prevents the rubber valve from opening to too great an extent. The centre of the rubber valve does not rise off its seat. This type of valve is often used in air pumps; it is unsuitable for cases where there is a considerable head of water above the valve.

The bucket valves in the illustration consist of a disc of vulcanised fibre protected on its back by a bronze disc. The fibre valve rises off its seat, and is closed by the pressure of water upon it, and by a spring. This is a good form of valve, as the fibre does not require to bend, and it does not strike such a heavy blow on its seat as is the case with a solid metallic valve.

The foot valves shown are of the Kinghorn pattern; this valve consists of three or more thin bronze discs; the discs, with the exception of the upper one, are perforated, but in such a manner that the perforations do not come opposite to one another, so that when the discs are close together no water passes, but when opened a slight distance the water can pass through the perforations. The lift of each disc is very small, and as they are thin and light, and as there is a small quantity of water between each disc when the plunger reverses its stroke,

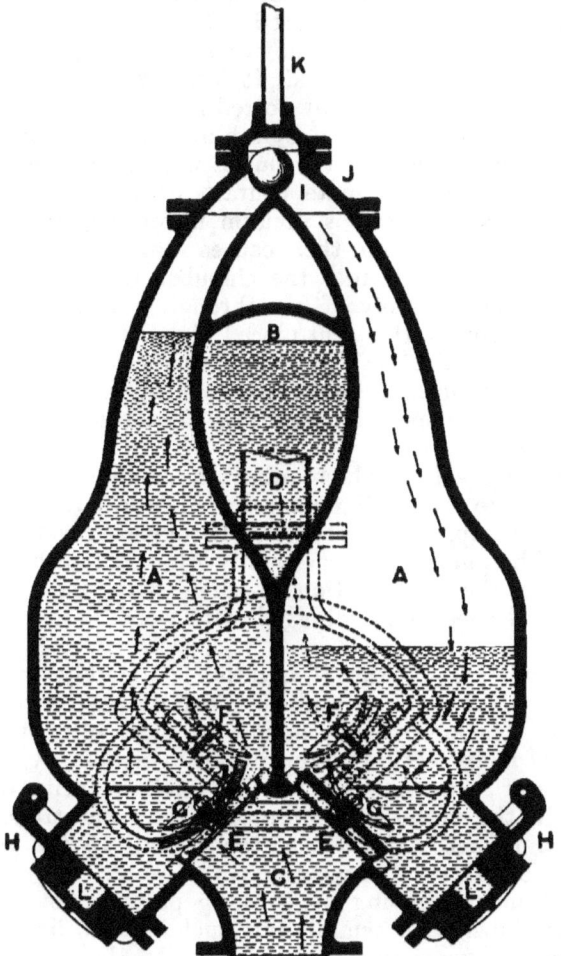

Fig. 99.—Pulsometer pump.

the valve closes very gently. Kinghorn valves are largely used in marine air pumps, and are found to last a long time.

The Pulsometer Pump.—This pump which is shown in section by Fig. 99 is of great simplicity, and is extremely useful in cases of emergency. The pump can be lowered down a pit by a chain, steam being supplied to it by a flexible or ordinary steam pipe. No foundations or fixings are required. The pump consists of two chambers A, A, which join at the neck J. In the neck is placed a single ball valve, I. The ball oscillates on its seat, so that while the entrance to one chamber is open, the entrance to the other is closed. The pump is first charged with water, the steam enters one of the chambers which happens to be left uncovered by the ball, it forces down the water until it reaches the outlet shown by dotted lines, when the surface of the water (previously covered by a film of air) is broken up, the steam then blows through the outlet with a certain amount of violence, comes into intimate contact with a large amount of cold water and condenses, a vacuum is formed, and the ball is pulled over. The vacuum then causes water to rise up through the suction valves E, E, into the chamber just emptied; it also draws in a small supply of air through a snifting valve not shown by the illustration, but placed high up in the narrow part of the chamber. *The air thus drawn in forms a cushion between the steam and the water on the succeeding stroke.* While this has been going on, the operation first described has been proceeding in the second chamber.

The snifting valves which admit a small supply of air at every stroke are seldom if ever referred to in published descriptions of the pump, or by lecturers, yet they are a very important feature. The pump, it is true, will work without them, but the difference in the consumption of steam when these valves are removed, and their holes plugged with wood, is at once noticeable.

The chamber B in the illustration is merely an air-vessel connected to the suction, and does not affect the principle of the pump. The air-vessel is similar to those often fitted to the discharge of reciprocating pumps for ensuring a fairly continuous flow of water; at every stroke the small quantity of air imprisoned in the vessel is compressed; between strokes the air expands and forces out some of the water. In the pulsometer the air-vessel ensures a regular supply of water.

The pulsometer having no working parts is not affected by grit; it will deal with semi-liquids such as mud, liquid cement, sewage-sludge, &c. It requires no oil; the steam used is condensed, and is therefore not a nuisance when the pump is worked down a pit or in a confined space. The pulsometer will raise water to a height of from 70 to 80 feet; it should be placed within from 6 to 15 feet of the surface of the water, depending upon the size of the pump. The only objection to this form of pump is that its consumption of steam is higher than that of a good reciprocating pump. The patents in connection with the pulsometer having expired, a good many pumps on similar lines are now made, such as the Aqua-thruster, the Expulsor pump, and others. Whether these pumps are as good as the original pulsometer or not the author is unable to say; he has not had one under his own observation.

An ingenious device for enabling a large quantity of water at a low level to raise a small quantity to a higher level is called a **Hydraulic Ram.** The action is this:—Water flows down a long pipe and is allowed to run to waste through a valve which is kept open by a weak spring; when the water has attained a sufficient velocity it closes the valve, and forces open a second valve, which admits to the pipe leading to the higher level; the kinetic energy acquired by the water is sufficiently great to carry a small quantity to the higher level. When the velocity of the water falls off, the waste valve again opens, and the cycle is repeated.

The materials of which pumps should be made for pumping special liquids require to be carefully considered. The following have worked well:—

Amoniacal liquor,	Cast iron entirely.
Naphtha,	,, ,,
Tar and creosote,	,, ,,
Petroleum,	Cast iron and brass.
Weak acids,	Gun-metal.
Sugar, treacle, and malt,	,,
Vinegar,	Lead.
Salt water,	Copper 88, tin 10, zinc 2.

Flow of Water in Long Pipes.—A formula which gives the head necessary to overcome the friction of water in pipes is sure to be very useful to students. This formula enables an engineer to decide whether a pipe of a given size is, or is not, large enough to pass a certain quantity of water in a given time.

The formula, which is given in Box's useful book on hydraulics, is as follows:—

$$H = \frac{G^2 \times L}{(3D)^5};$$

where G = gallons per minute.
L = length of pipe in yards.
D = diameter of pipe in inches.
H = head in feet.

As a table of the fifth powers of numbers is not always

available, and the formula is such a useful one, a table of fifth powers is appended:—

TABLE XX.—FIFTH POWER OF NUMBERS.

No.	5th Power.	No.	5th Power.	No.	5th Power.
1	1	34	45,435,424	67	1,350,125,107
2	32	35	52,521,875	68	1,453,933,568
3	243	36	60,466,176	69	1,564,031,349
4	1,024	37	69,343,957	70	1,680,700,000
5	3,125	38	79,235,168	71	1,804,229,351
6	7,776	39	90,224,199	72	1,934,917,632
7	16,807	40	102,400,000	73	2,073,071,593
8	32,768	41	115,856,201	74	2,219,006,624
9	59,094	42	130,691,232	75	2,373,046,875
10	100,000	43	147,008,443	76	2,535,525,376
11	161,051	44	164,916,222	77	2,706,784,157
12	248,832	45	184,528,125	78	2,887,174,368
13	371,293	46	205,962,976	79	3,077,056,399
14	537,824	47	229,345,007	80	3,276,800,000
15	759,375	48	254,803,968	81	3,486,784,401
16	1,048,576	49	282,475,249	82	3,707,398,432
17	1,419,857	50	312,500,000	83	3,939,040,643
18	1,889,568	51	345,025,251	84	4,182,119,424
19	2,476,099	52	380,204,032	85	4,437,053,125
20	3,200,000	53	418,195,493	86	4,704,270,176
21	4,084,101	54	459,165,024	87	4,984,209,207
22	5,153,632	55	503,284,375	88	5,277,319,168
23	6,436,343	56	550,731,776	89	5,584,059,449
24	7,962,624	57	601,692,057	90	5,904,900,000
25	9,765,624	58	656,356,768	91	6,240,321,451
26	11,881,376	59	714,924,299	92	6,590,815,232
27	14,348,907	60	777,600,000	93	6,956,883,693
28	17,210,368	61	844,596,301	94	7,339,040,224
29	20,511,149	62	916,132,832	95	7,737,809,375
30	24,300,000	63	992,436,543	96	8,153,726,976
31	28,629,151	64	1,073,741,824	97	8,587,340,257
32	33,554,432	65	1,160,290,625	98	9,039,207,968
33	39,135,393	66	1,252,332,576	99	9,509,900,499

Example.—Suppose we have a pipe 80 yards long and only 6 inches in diameter, and we wish to pass 1,000 gallons of water per minute through it, what head or pressure will be required to make this quantity of water flow through the pipe? The calculation will be as follows:—

$$\frac{1,000^2 \times 80}{(3 \times 6)^5}.$$

The square of 1,000 is 1,000,000, and multiplying this by 80 we get:—

$$\frac{80,000,000}{(18)^5}.$$

The fifth power of 18 is 1,889,568, so we get—

$$\frac{80,000,000}{1,889,568} = \text{say } 42 \text{ feet.}$$

The answer shows that a head of 42 feet would be required to force 1,000 gallons of water through a pipe 6 inches in diameter and 80 yards long. Such a head would be quite inadmissible under ordinary conditions, and the calculation shows, either that the quantity of water would have to be reduced or a larger pipe used.

Now let us try the effect of passing the same quantity of water per minute through a pipe 12 inches in diameter.

$$\frac{1,000^2 \times 80}{(3 \times 12)^5},$$

or $\dfrac{80,000,000}{(36)^5}$,

or $\dfrac{80,000,000}{60,466,176} = 1\cdot 33$ feet.

A head of 1·33 feet, corresponding with a pressure of about ·57 lb. per square inch, would be a reasonable one to allow for forcing the water through the pipe, assuming the water has to be pumped; if, on the other hand, a natural head of between 5 and 6 feet is available, the reader will find that according to the above formula a pipe 9 inches in diameter is sufficiently large.

USEFUL HYDRAULIC MEMORANDA.

1 cubic foot of fresh water at 32° F. weighs 62·418 lbs.
1   „   „   *39·1°–40°   „   62·425 „
1   „   „   60°   „   62·321 „
1   „   „   100°   „   62·022 „
1   „   „   200°   „   60·081 „
1 gallon of fresh water at   60°   „   10   „
1   „   = 277·27 cubic inches.
1 lb. of water at 32° measures 27·68 cubic inches.
1   „   60°   „   27·72   „

A column of water at 40° F. 1 foot high = ·4335 lb. per sq. in.
     „   „   60°   „   = ·4328   „

The capacity of a cylinder in gallons $= \dfrac{A \times L}{277\cdot 27}$;

where A = area of cylinder in inches.
L = length   „   „

Pressure of water at 40° F. on the side of a vessel
    $= A \times D \times 62\cdot 425$ lbs.;

where A = area of side in feet.
D = half depth in feet.

* Water is at its greatest density at a temperature of 39·1°.

CHAPTER XIII.

GAS AND OIL ENGINES.

SUCTION GAS PLANT.

In gas and oil engines the fuel is burnt in the cylinders, hence they are called internal combustion engines, as distinguished from external combustion engines, such as those driven by steam, in which the fuel is burnt in a furnace outside the engine.

Burning the fuel in the cylinder enables a higher thermal efficiency to be obtained than is possible with an external combustion engine, but introduces difficulties which are absent in the latter. The advantages and disadvantages of the gas engine will be discussed later. In the meantime the principles upon which gas engines usually work will be described.

Otto Cycle.—In the great majority of gas and oil engines the Otto cycle or four-stroke cycle is adopted; with this cycle one explosion is obtained during every two revolutions of the engine crank. The action, commencing with the explosion, is as follows :—

1. Explosion—Outward stroke.
2. Inward stroke—Piston drives out exhaust gases.
3. Outward stroke—Piston draws in mixture of gas and air.
4. Inward stroke—Piston compresses mixture.

When the cycle is again repeated.

Description of Engine.—Fig. 100 represents a 10 B.H.P. gas engine * stripped of all details, such as governor, lubricators, cams, levers, &c. The piston is shown at the beginning of its stroke, having compressed its charge ready for ignition. After the explosion has taken place, and the piston has reached the end of its stroke, the exhaust valve will be lifted by a lever worked from a cam on the shaft, a portion of which is shown at the front end of the engine. This shaft is driven by gearing, and only makes one revolution for every two of the engine. After the burnt gases have been driven out, the gas and air valves, which work horizontally, are opened by levers worked from cams on the shaft already referred to, and a charge of gas

---

* Constructed by the Railway and General Engineering Company of Nottingham.

240 MECHANICAL ENGINEERING FOR BEGINNERS.

and air is drawn in; this mixture is then compressed by the piston on its return stroke, when all is ready for ignition.

The engine illustrated has many good features. In the first place, the valves and their seatings can easily be removed for

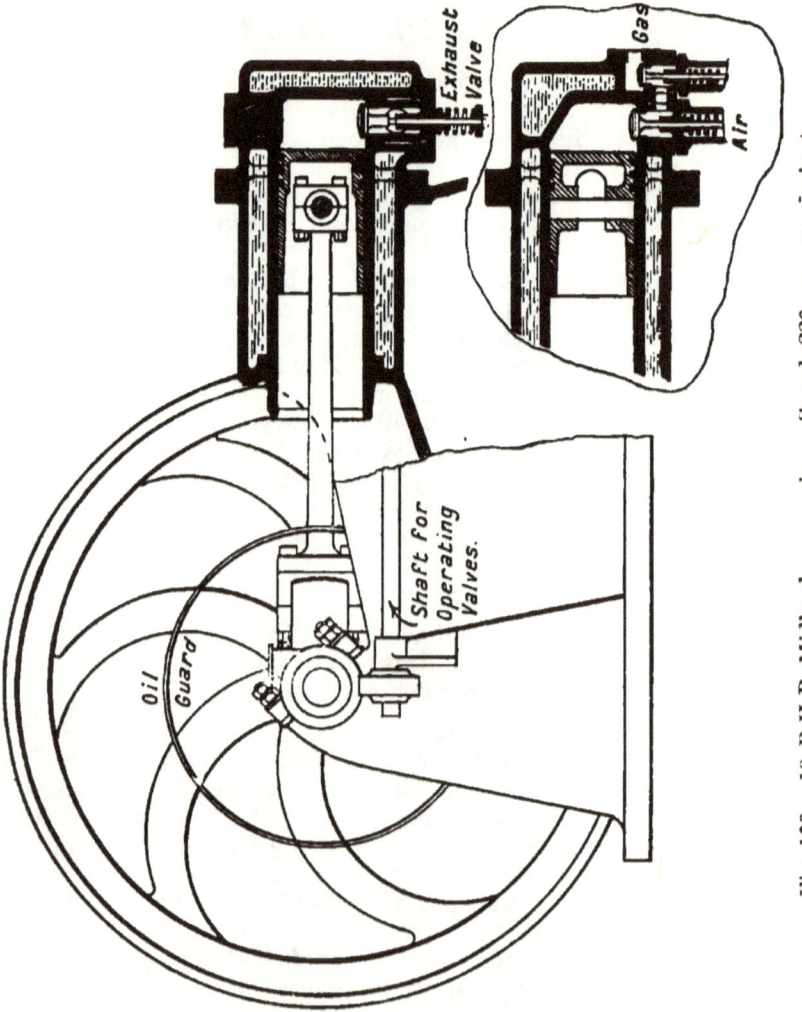

Fig. 100.—10 B.H.P. Midland gas engine. Speed, 220 revs. per minute.

examination and cleaning; in the second place, the cylinder and water-jacket are not overhung and carried entirely by the flange at one end, as is so commonly met with. It may be considered by some that valves working horizontally, as the gas and air valves in the engine illustrated, are not so good as those which have a vertical lift; in practice, however, the horizontal valves have been found to work well.

The points about a gas engine to which attention may be directed, taking them in the order of the cycle already given, are as follows:—

Methods of Ignition.—In the earlier Otto engines, in which a single slide valve was used for admitting the mixture and exhausting the burnt gas, the slide valve brought forward at the right moment a small pocketful of lighted gas, and placed it opposite a port at the end of the cylinder, the result being that the compressed mixture immediately exploded. The slide valve with its rubbing surfaces was found to be unsuitable for high speeds and for high temperatures and pressures, and was abandoned in favour of valves of the mushroom type, as shown by the illustration. When the mushroom valves were adopted, the tube method of ignition came into general use. This method consists of keeping a tube red hot by allowing a flame to play constantly upon its outside; when the mixture has been compressed, a small quantity is admitted to the inside of the tube, and ignition takes place. This method of ignition is still largely used, but in large gas engines, and in many modern small gas and oil engines, the electric method of ignition is employed.

Electric ignition may be carried out in several ways. In the majority of cases a small dynamo, called a magneto, is used. If the current produced by the magneto is of low tension—the type usually employed with gas engines—then the current is suddenly interrupted in a place which is accessible to the explosive mixture; this sudden breaking of the circuit causes a spark which fires the mixture. If the magneto is of the high-tension type, as frequently used with petrol engines, the circuit is made and broken outside the cylinder, the spark inside the cylinder or explosion chamber being caused by the current jumping across the fixed points of a sparking plug screwed into the cylinder.

When a magneto is not used, a 2-cell accumulator giving a current at an E.M.F. of about 4 volts is generally employed, and an induction coil is used to increase the E.M.F. of the current, so as to enable it to jump across the points of the sparking plug. The current in such cases is interrupted by means of a trembler placed at the top of the induction coil. Such accumulators require to be charged periodically, and are much more troublesome than a magneto.

In the Diesel oil engine no ignition apparatus of any kind is used; the air is compressed to a pressure of about 500 lbs. per square inch, corresponding with a temperature of about 1,000° F., which temperature is sufficiently high to fire the fuel immediately it is injected.

Exhausting and Scavenging.—The mere fact of opening the exhaust valve is not sufficient to ensure that all the burnt gases will leave the cylinder; in fact, the clearances will remain filled with them. These remaining products of combustion reduce the value of the next incoming explosive mixture, and, if any incandescent particles are left in the cylinder, they may cause pre-ignition of the charge. To ensure the removal of the burnt gases various expedients, called scavenging, are adopted. The simplest is that usually employed in connection with the Crossley engine; it consists of an exhaust pipe of suitable diameter 60 or 65 feet long. The exhaust gases travel along this pipe at such a speed that (to express it colloquially) they find it difficult to stop, and actually create a vacuum in the cylinder.

This arrangement is not altogether satisfactory at light loads, and in large gas engines it is not practicable to arrange for a suitable length and diameter of exhaust pipe. Air pumps are therefore frequently used for forcing air into the cylinders and thus scavenging them before the fresh charge is admitted. The air pumps are driven by some reciprocating part of the engine or by an auxiliary crank.

Drawing in Mixture of Gas and Air and Governing.—A gas engine may be governed in three ways—(1) By omitting to open the gas inlet valve when the speed of the engine rises beyond a certain point; this is called governing on the hit-and-miss principle, the miss occurring when the speed is too high. (2) By altering the mixture of gas and air. (3) By altering the quantity of the mixture admitted without altering the quality. The first method is that usually adopted in small and medium powered gas engines; it is simple, but the governing is not very accurate unless a really heavy flywheel is used. Governing by either the second or third methods is preferable—viz., by altering the quality or quantity of the mixture. By reducing the quantity of the mixture admitted to the cylinder, the compression is reduced and the full value of the explosive mixture is not obtained. On the other hand, by reducing the amount of gas admitted and so altering the quality of the mixture, the gas is not burnt to the best advantage. Both systems have their advocates.

Compressing the Mixture.—It has been conclusively proved that a high degree of compression results in a high explosive force, and *vice versâ*; a mixture which is highly compressed burns more quickly than one which is only slightly compressed. Consequently, during recent years the clearances in gas engines have been reduced, and the compression increased. The amount to which the charge can safely be compressed without risk of pre-ignition, depends upon the scavenging and water-cooling arrangements.

In modern gas engines of large power, and even in small engines in which economy is studied, the compression ranges from 150 to 200 lbs. per square inch. In the Diesel engine, the air is compressed to about 500 lbs.; the fuel is then injected by a blast of air at a pressure of about 600 to 650 lbs., when the mixture immediately ignites.

We have said that the majority of gas engines work on the

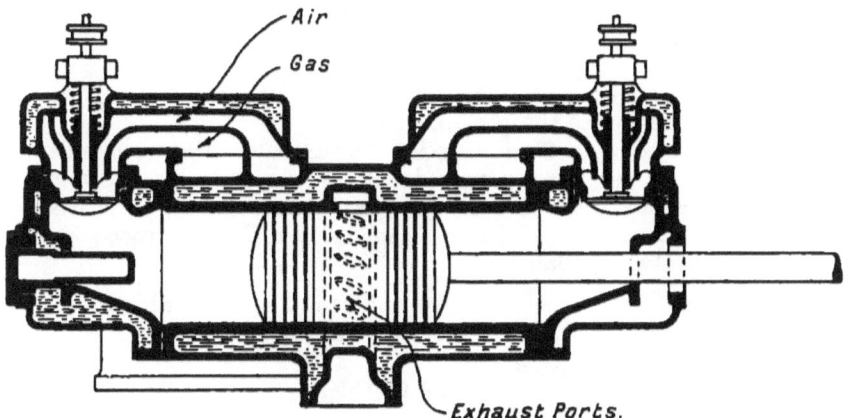

Fig. 101.—Körting gas engine.

Otto cycle; the most noteworthy exception is the Körting engine, shown by Fig. 101. This engine is double-acting, and there are no exhaust valves. The exhaust ports are placed in the middle of the cylinder, as shown by dotted lines; when the piston reaches the end of its stroke in either direction, it uncovers the exhaust ports; as soon as the ports are uncovered, a charge of fresh air is blown in, and the burnt gases are expelled. When the piston has moved sufficiently far to close the exhaust ports, a supply of gas is pumped in, the air and gas inlet valve then closes, and the mixture is compressed.

By this arrangement there is an impulse every outward stroke, and as the engine is double-acting, there are two impulses per revolution. Such an engine occupies much less space than a single-cylinder engine, in which there is only one impulse during every two revolutions. The gas and air pumps are placed alongside the cylinder, and are worked off an auxiliary crank; they are not shown in the illustration.

Double-acting engines, working on the four-stroke cycle, are frequently made on the Continent. These engines occupy less space than those of the single-acting type, but the parts are not quite so accessible for cleaning; moreover, a gland is required at the piston-rod end of the cylinder. It will be noticed that in the single-acting engine, illustrated by Fig. 100, one side of the piston or trunk is open to the air, and no gland is required; the piston can be withdrawn by undoing the bolts of the connecting-rod brasses. These are good features. The gland of a double-acting gas engine, unless designed with great care, and water-cooled, is likely to be a source of trouble.

In double-acting engines, and in single-acting engines when the power exceeds 250 B.H.P., it is advisable to water-cool the piston and rod; water for this purpose is admitted to the crosshead by means of swinging link pipes, and thence through the hollow piston-rod to the piston.

Amount of Gas Consumed.—The amount of gas consumed in an engine per horse-power developed, depends largely upon the richness or otherwise of the gas. The number of British thermal units contained in gases made by different processes is approximately as follows :—

Town gas,	600 to 680	B.T.U. per cubic foot.
Dowson gas,	150 to 164	,, ,,
Producer gas (suction gas plant),	} 125 to 155	,, ,,
Mond gas,		
Blast-furnace gas,	100 to 120	,, ,,

Town gas requires to be mixed with about 10 times its volume of air, while producer gas requires about $1\frac{1}{4}$ or $1\frac{1}{2}$ times its volume of air, in order to make the best explosive mixture in the engine. The consumption of rich town gas in an engine of, say, 20 H.P. should be between 14 and 16 cubic feet per I.H.P. per hour, which is equivalent to between $16\frac{1}{2}$ and 19 cubic feet per B.H.P. The consumption of producer gas containing 130 or 140 B.T.U. in a good engine should not exceed about 60 cubic feet per I.H.P. per hour, or 70 cubic feet per B.H.P.

Horse-power of Gas Engines.—The indicated horse-power of a gas engine, working on the Otto or any other cycle, may be found, provided the mean pressure in the cylinder is known, by the following formula :—

$$\frac{S \times N \times A \times P}{33,000};$$

where S = stroke of the piston in feet.
N = number of explosions per minute.
A = area of the piston in inches.
P = mean pressure exerted on the piston during the stroke.

Example.—What is the indicated horse-power of a gas engine having a cylinder 10 inches in diameter, 18 inches stroke, and speed 190 revs. per minute, the mean effective pressure on the piston being 70 lbs.? The calculation is

$$\frac{1 \cdot 5 \times 95 \times 78 \cdot 5 \times 70}{33,000} = 23 \cdot 7 \text{ I.H.P.}$$

The above calculation is for an engine working on the Otto cycle, in which there is an explosion for every 2 revs. of the engine.

The actual horse-power of a gas engine can, however, only be determined with accuracy by a brake trial, or by coupling it to a dynamo, the efficiency of which is known, for, in a gas engine, the indicator cards which purport to give the mean pressure are not altogether to be relied upon.

The maximum temperature attained in the interior of a gas-engine cylinder by using a mixture of 1 part of town gas to 9 parts of air, and compressing it to about 100 lbs. pressure, is between 3,000° and 3,500° F. The higher of these temperatures is above the melting point of platinum, so that the necessity of a water-jacket is apparent. The maximum pressure reached in the cylinder is about 3 or $3\frac{1}{2}$ times the compression pressure.

Thermal Efficiency.—The thermal efficiency, either per indicated or per brake horse-power, of a gas engine is easily ascertained if the number of cubic feet of gas used per horse-power per hour, and the calorific value of the gas, are known. Let us find the thermal efficiency per B.H.P. of a large gas engine using 60 cubic feet of producer gas per I.H.P. per hour. If the mechanical efficiency of the engine is between $85\frac{1}{2}$ and 86 per cent., the consumption of gas will be 70 cubic feet per B.H.P. per hour. We will assume that the calorific value of the gas is 130 B.T.U. per cubic foot. The heat used by the engine per B.H.P. is, therefore, 70 × 130 = 9,100 B.T.U. per hour. We read in Chapter III. that a B.T.U. = 772 ft.-lbs., and a H.P. 33,000 ft.-lbs; therefore, 42·75 B.T.U. per minute, or 2,565 B.T.U. per hour, are the equivalent of 1 H.P. The engine uses 9,100 B.T.U. per brake horse-power, so that the thermal efficiency of the engine is $\frac{2,565}{9,100} = \cdot 28$. The absolute thermal efficiency of the engine per B.H.P. is therefore ·28.

If the engine only uses ·28 of the heat supplied to it, the question which naturally arises is — What becomes of the remainder of the heat? The distribution of the heat originally contained in the gas is approximately as follows :—

·38 carried off in the exhaust gases.
·29 carried off by the jacket cooling water by radiation.
·28 given up as brake horse-power.
·05 used in overcoming the friction of the engine.
———
1·00

Although a thermal efficiency of ·28 is apparently low, it compares favourably with that of a steam engine. The absolute thermal efficiency of a triple-expansion condensing engine, working with saturated steam at 160 lbs. pressure, and using 14 lbs. of steam per B.H.P., is only ·162.

Speeds of Gas Engines.—The speeds at which gas engines usually run are approximately as follows :—

5 B.H.P.,	250	revs. per minute.		
10	„	220	„	„
20	„	200	„	„
50	„	175	„	„
100	„	160	„	„
500	„	120	„	„

Suction Gas Plant.—The history of the modern suction plant is briefly as follows:—It was found by the late Mr. Siemens that it was more economical and convenient to turn cheap fuel into gas, and then burn the gas in regenerative furnaces, than to utilise the heat derivable from complete combustion of the coal in the first place. To turn the coal into gas it was placed on a deep layer on a grate and partially burnt, the air that found its way through the coal not being of sufficient quantity to allow of complete combustion. The gas that was given off consisted chiefly of carbon monoxide (CO) and nitrogen, together with a small quantity of hydrogen. The gas was of low calorific value, and, in its production, not less than 30 per cent. of the heat contained in the coal was lost. The small quantity of hydrogen, too, which it contained prevented the rapid ignition which is necessary in an explosive mixture. In order to improve the quality of the gas, Mr. Dowson, in 1878, turned a jet of steam (H_2O) into the hot fuel, the result being that the hydrogen was liberated and was added to the gas, while the oxygen combined with the carbon. The temperature of the furnace was reduced and the quality of the gas improved, the calorific value of the Dowson gas being within 15 per cent. of that of the solid coal.

Although gas produced on the Dowson system was of fair calorific value and inexpensive, there were at least two reasons why the system did not come into general use for driving gas engines. The first was the high cost of the plant, and the second the great amount of floor space required. A boiler, working at about 50 lbs. pressure, was required to drive the steam and air through the coal, and a gas-holder was provided to contain the gas after it had been produced.

In a modern suction gas plant the heat generated in the producer is used to form the steam, and, as the heat is not sufficiently great to generate steam at a pressure sufficiently high to force it through the fuel, the outward stroke of the engine is used to suck the steam and air through the fire. A gas-holder, if placed between the generator and the engine, would do harm, as the effects of the suction stroke would not be felt in the generator; hence both boiler and gas-holder have been done away with.

One of the simplest forms of suction gas plants* now made is shown by Fig. 102. The generator or producer A in which the fuel is burnt is surrounded by a jacket or vaporiser, B; this jacket is filled with coke or any similar material; the material is kept moist by sprinkling water upon it, and is warmed by the heat of the producer. The air required for combustion is compelled to pass through the belt of wet coke in the jacket, and in doing so becomes charged with moisture; the suction stroke of the engine draws the hot moist air through the fuel, as already explained. The amount of water sprinkled on to the coke in the jacket is regulated automatically by the suction of the engine. After the gas has been made in the producer, it is necessary to pass it through the scrubber shown on the right-hand side of the illustration. The scrubber consists merely of a cylinder filled with coke, through which a stream of water is kept flowing in a downward direction; this water carries away any tar suspended in the gas. After leaving the scrubber the gas is usually taken to a box, which helps to reduce any violent fluctuations of pressure, and is then taken to the engine. The fan shown on the left side of the illustration is only used at starting, when the cock C is opened, and the gases pass away to the atmosphere.

Amount of Gas made and Power developed per Pound of Coal.—A producer similar to the one illustrated will make about 80 cubic feet of gas for every pound of anthracite coal consumed, and as a good engine of fairly large size uses only

* Constructed by the Dowson Gas Co. of Westminster.

about 70 cubic feet of producer gas per B.H.P., it follows that 1 lb. of anthracite will give about 1·14 B.H.P. It is the

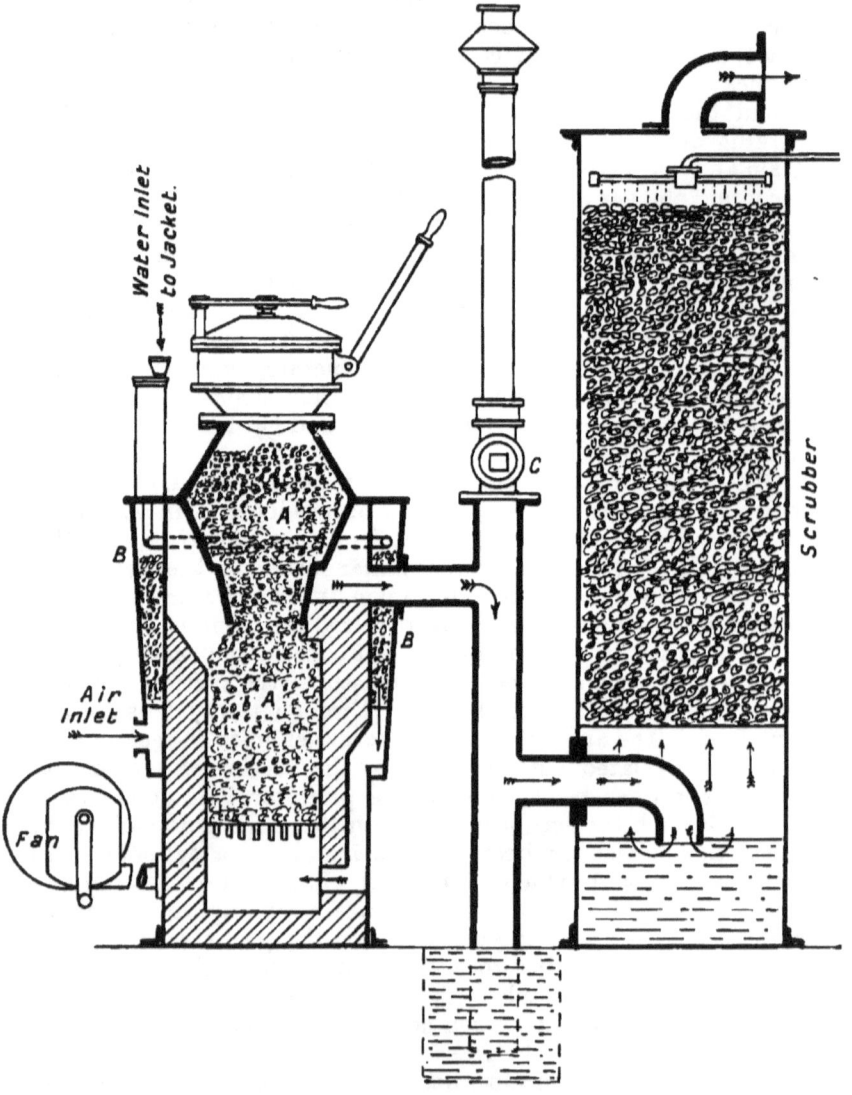

Fig. 102.—Suction gas producer.

practice of many makers of suction gas plants to guarantee that ·9 lb. of anthracite will give 1 B.H.P., but at some trials of several 15 to 20 B.H.P. engines working with suction gas, carried out by the Royal Agricultural Society in 1906, the consumption was a little higher. The lowest consumption was 1·04 lbs. of anthracite, and the highest 1·47 lbs., the average consumption of the eleven engines being 1·21 lbs. of anthracite per B.H.P. The consumption of an engine working with a Dowson producer, as illustrated, was, at the trials in question, 1·09 lbs. per B.H.P. The average consumption of eleven plants working with coke was 1·4 lbs. per B.H.P.

Water required for Suction Gas Plants.—It is usually reckoned that about 1 gallon of water is required per B.H.P., nine-tenths of which is used in the scrubber, the remaining tenth in the generator. In the trials carried out by the Royal Agricultural Society, the consumption of water was considerably higher than 1 gallon. The average consumption of twelve plants was 1·89 gallons of water per B.H.P., or about three-fourths as much as would be used by a good non-condensing steam engine. The amount of water used by the plant which secured the gold medal was 1·14 gallons per B.H.P.

Mond Gas.—Bituminous coal cannot be used in a suction gas producer, only anthracite coal or coke. Dr. Mond, however, has introduced a process by which cheap bituminous coal may be used for the production of gas. The Mond process consists of introducing an enormous quantity of superheated steam and air to the generator. The greater portion of the steam passes out of the producer undecomposed; it is then condensed, and its heat is utilised. The temperature of the generator is kept very low by the introduction of the very large quantity of steam and air, and a large amount of sulphate of ammonia is obtained; this forms a valuable by-product. No tar is produced. This system, like that first introduced by Mr. Dowson, requires a separate boiler. About 70 cubic feet of gas are produced from 1 lb. of bituminous coal.

The analyses of gases produced by the Dowson, Suction, and Mond processes are approximately as follows:—

	Dowson System.	Suction Gas System.	Mond System.
Hydrogen (H),	19·8	17·5 to 20·5	24·8
Oxygen (O),	...	0·5 ,, 0·2	...
Carbon monoxide (CO),	23·8	18·5 ,, 21·5	13·2
,, dioxide (CO_2),	6·3	7·0 ,, 7·5	12·9
Marsh gas (CH_4),	1·3	1·5 ,, ·5	2·3
Nitrogen (N),	48·8	55·0 ,, 49·8	46·8
	100·0	100·0 100·0	100·0

Starting Gas Engines.—Small gas engines may be started by the driver pulling round the flywheel by hand, and thus drawing in the explosive mixture and compressing it; but in the case of large engines this is impracticable, and special means for starting have to be provided. Some engines, such as the Westinghouse gas engine and the Diesel oil engine, are started by means of compressed air; the engine is provided with a pump which compresses a certain quantity of air before stopping, and forces it into a drum provided for the purpose. Other engines are provided with a starter, consisting of a chamber into which a mixture of gas and air is pumped by hand; the flywheel of the engine is then barred round by a lever with a pawl at the end, teeth being cast in the flywheel for the purpose. When the piston is in the right position, the cylinder is placed in communication with the starter, the mixture is fired, and the resulting explosion in the cylinder is sufficient to start the engine. Other engines merely have a pinion with large hand wheel; the pinion works in mesh with teeth in the large flywheel; the large hand-wheel is constructed on the free-wheel principle, so that it is not driven round when the gas engine is running.

Comparative Merits of Steam, Gas, and Oil Engines.— The cost of fuel per horse-power for gas, steam, oil, and petrol engines is approximately as follows :—

	Penny per B.H.P. per hour.
Gas engine using producer gas, and consuming $1\frac{1}{4}$ lbs. of coke per B.H.P. The cost of coke is taken at 15s. per ton,	·1
Gas engine using 20 cubic feet of town gas per B.H.P. The cost of town gas is taken at 2s. 6d. per 1,000 feet,	·6
Condensing steam engine using $1\frac{3}{4}$ lbs. of coal per B.H.P. The cost of coal is taken at 15s. per ton,	·14
Non-condensing steam engine using 3·75 lbs. of coal per B.H.P. The cost of coal is taken at 15s. per ton,	·3
Oil engine using ·085 of a gallon, at 4d. per gallon,	·34
Petrol engine using ·11 of a gallon, at 1s. 4d. per gallon,	1·76

The cost of coal per ton, of gas per 1,000 cubic feet, and of oil and petrol per gallon are given so that the reader can make any necessary correction in the comparative costs to suit the cost of fuel in his own locality.

If the water required for either steam or gas plant has to be paid for, the cost must be added to the figures given.

It will be seen that, so far as fuel alone is concerned, a gas engine using producer gas is the cheapest form of motive power, also that a gas engine using town gas at 2s. 6d. per 1,000 feet is, with the exception of a petrol engine, the most expensive of those given. Fuel, however, is not the only item to be taken into consideration; there is, for instance, the cost of attendance. A small gas engine running on town gas will run without attention, whereas a steam boiler requires a man to stoke it and to see that the water level is maintained. If we take the case of a 10 horse-power gas engine running fifty-four hours per week on town gas costing ·6 of a penny per B.H.P. without attendance, and a steam engine costing only ·3 of a penny per B.H.P., but necessitating 18s. per week being spent in wages, the comparison will come out in favour of the gas engine. It is largely due to the fact that small gas engines can be run practically without attention, and can be stopped and started without any fuel being used while the engine is standing, that they are so widely used in small works. With large gas engines and producers the cost of attendance is as great as, if not greater than, that required with steam plant.

On grounds other than those of running costs and possible risk of boiler explosions, the advantages are all on the side of the steam engine. Ease of starting, silence, general sweetness of running, and reliability, are qualities much more marked in the steam engine than in its rival. The author a few years ago spent some hours in an electric generating station where gas engines were employed, and he subsequently revisited the same station when the gas engines had been replaced by steam plant. The strongest impressions produced at the first visit were the terrible noise (it was quite impossible to hear oneself speak in the engine-room), and the pallid appearance and anxious look upon the faces of the attendants. The chief engineer said that he was much handicapped by illness amongst his staff due to noxious fumes.

On the second visit, when the generators were driven direct by 600 H.P. single-acting engines, the whole conditions were changed; one could converse, without raising one's voice, in any part of the engine-room, and the attendants all looked well; and while during the first visit one's chief desire was to escape as quickly as politeness permitted from what was almost an inferno, on the second visit one was tempted to linger unduly amongst the silent-running plant, and around the switch board.

We have said that the steam engine is more reliable than the gas engine. Perhaps this statement should be qualified in this way. It is not suggested that a gas engine, provided its valves are kept clean, and supplied with a proper explosive mixture, is not reliable, but the author's meaning is this—given water, coal, and a sound boiler one can be sure of getting steam of some definite and easily ascertainable pressure, and if this steam fails to drive the engine the fault can be located easily. But in the case of a producer plant, given coal, water, and a producer, one cannot be absolutely sure of getting a gas of some definite explosive strength. On paper it is, of course, a very easy matter, but in practice the strength of the explosive mixture varies considerably, and unless the engine is of sufficient size to give the full power required with a weak mixture, its speed will fall off and the whole work of the factory may be deranged. A small gas engine working with town gas does not experience great variations in the quality of the gas supplied, and is probably as reliable as a steam engine.

Oil Engines work on the same principles as gas engines; but the former, with the exception of the Diesel engine, require some form of vaporiser to turn the heavy oils used into vapour. The difference between the various makes of oil engines consists chiefly in the vaporiser. In the Hornsby engine, as shown by the illustration, this forms a continuation of the cylinder, and when compression is completed the temperature is sufficiently high to fire the mixture without any ignition apparatus. A lamp giving out great heat is used at starting, and is retained until the vaporiser has become sufficiently warmed to vaporise the oil.

In the Priestman engine the vaporiser is separate from the cylinder, and oil is vaporised by being passed through a fine jet and heated by the exhaust gases, while the mixture is fired by electric ignition

The characteristics of the Diesel engine have been referred to in the remarks on ignition and compression.

Petrol Engines.—Petrol engines work on precisely the same principles as gas engines. The four-stroke cycle is generally adopted in this country and abroad. In the United States, however, a certain number of cheap two-stroke cycle engines are made for launches. Petrol, owing to its light character, vaporises very easily; the vaporisation is effected in a chamber called a carburetter. The petrol is drawn up in the form of spray through several fine openings by the suction of the engine

GAS AND OIL ENGINES. 253

and mixes with a current of air also drawn up by the engine suction.

Consumption of Oil and Petrol.—The amount of oil used in a good oil engine is about ·085 of a gallon or ·7 of a lb. per B.H.P. In the Diesel engine the consumption is as low as ·41 lb. in an engine of 100 H.P., but the first cost of this engine is rather high. The cost of suitable Russian oil "naked on wharf" is about 45s. a ton. The cost of barrels and carriage may bring the cost up to £4 per ton, or, roughly, fourpence per gallon. The consumption of petrol is about ·1 to ·125 of a gallon, or ·7 to ·9 of a lb. per B.H.P. per hour.

Specific Gravity of Oils.—We have spoken of heavy and light oils; the specific gravity of crude petroleum is about 0·92 so that a gallon weighs about 9·2 lbs. The ordinary household petroleum used for lamps weighs about 8·2 lbs. per gallon (specific gravity 0·82); while petrol or petroleum spirit weighs from 6·6 to 7·2 lbs. per gallon (specific gravity 0·66 to 0·72).

One lb. of crude petroleum contains about 20,000 B.T.U., while 1 lb. of petrol contains about 17,500 B.T.U.

Flash Point.—The point at which an oil will vaporise when heated is termed its flash point—*i.e.*, if an oil vaporises at 75° F. its flash point is said to be 75°. An oil, the flash point of which is below 73°, is not considered legally safe in this country, and may not be stored without restrictions; efforts are being made to raise this flash point on account of the numerous accidents which occur with cheap lighting oils. The flash point of heavy petroleum, such as is used in oil engines, is about 95°.

CHAPTER XIV.

STRENGTH OF BEAMS AND USEFUL INFORMATION.

Strength of Beams.—To describe elaborate methods for calculating with extreme accuracy the strength of beams of unusual section, either graphically or mathematically, would be beyond the scope of this book, and, in point of fact, such calculations are seldom made by the average draughtsman or designer, unless engaged upon bridge or crane work. It is, however, necessary for every young engineer to be able to ascertain quickly and with fair accuracy, what stress is set up in the flanges, or body of a beam or cantilever when loaded in a given manner.

In those cases where the metal is distributed in the most efficient manner to resist bending—*i.e.*, in the form of an upper and lower flange with a web joining the two flanges, as in the case of a rolled steel joist, the stress in each flange can be found very quickly and simply by the formulæ given in column 1 of Table XXI., and, if the area of the flange is known, the stress per square inch is quickly ascertained. When the beam is supported at both ends and loaded, the upper flange is in compression, the lower flange in tension; and *vice versâ* when the beam is supported at one end only. The web is assumed to give stiffness only when this method of calculation is adopted.

In cases where the beam is of solid rectangular or of round section, the stress cannot be found so simply, as the relative value of the metal must be considered; for it is evident that metal situated midway between the top and bottom of the beam, or at its neutral axis, is not so efficient as metal placed at the greatest distance from the neutral axis. The breaking strength of rectangular beams can, however, be found in the manner indicated in the second column, and of beams of other sections by substituting for $B \times D^2$ the values for these sections given by the figures accompanying the illustrations.

It will be noticed in the case of solid rectangular beams, column 2, that the square of the depth is taken; the reason for this being that the strength of a rectangular beam varies as the square of its depth. It is obvious, however, that if the depth is squared and multiplied by the breadth, the number of inches so found, multiplied by the safe or breaking stress of the metal,

TABLE XXI.

	1. Stress in Each Flange in Beams of I Section (depth measured from centre to centre of flange).	2. Breaking Weight, in Cwts., of Plain Rectangular Beams.	Greatest Shearing Force.
(cantilever, end load)	$=\dfrac{WL}{D}$	$\dfrac{BD^2K}{L}$	W
(cantilever, distributed)	$=\dfrac{WL}{2D}$	$\dfrac{2BD^2K}{L}$	W
(simply supported, centre load)	$=\dfrac{WL}{4D}$	$\dfrac{4BD^2K}{L}$	$\dfrac{W}{2}$
(simply supported, distributed)	$=\dfrac{WL}{8D}$	$\dfrac{8BD^2K}{L}$	$\dfrac{W}{2}$
(fixed ends, centre load)	$=\dfrac{WL}{8D}$	$\dfrac{8BD^2K}{L}$	$\dfrac{W}{2}$
(fixed ends, distributed)	$=\dfrac{WL}{12D}$	$\dfrac{12BD^2K}{L}$	$\dfrac{W}{2}$
(off-centre load)	$=\dfrac{WL^1L^2}{LD}$	$\dfrac{BD^2L}{L_1 L_2}$	$\dfrac{W}{2}$
(solid circle)	$4\cdot 7 \times R^3$		

(hollow circle)	$4\cdot 7 \times \left(\dfrac{R_4 - r_4}{R}\right)$	W = weight applied.
		L = length of beam in ins.
		B = breadth of beam in ins.
		D = depth of beam in ins.
		K = coefficient.
		    = 65 to 90 cwts. for cast-iron test pieces free from blowholes.
		    = 45 to 50 cwts. for ordinary cast-iron structures.
		    = 90 to 95 cwts. for rolled steel joists.	

(I-section)	$\dfrac{BD^3 - 2bd^3}{D}$		

must not be taken as giving the load the beam will bear, coefficients for the ultimate strength of various materials are therefore used when this method of calculation is employed, and are given at the foot of the table.

The coefficient usually given for the strength of cast iron, although fairly applicable to ordinary cast structures, is too low, provided the metal is sound, free from blowholes, and free from internal stresses set up in cooling. For instance, we know that it is customary to specify that a bar of cast iron, 2 inches deep 1 inch wide, placed on supports 36 inches apart, shall carry a load of 30 cwts. suspended from its centre, and occasionally it is specified that such a bar of cast iron intended for cylinder liners shall carry a load of 40 cwts., and these results are obtained in actual practice. Now, if we work by the formula $\frac{4B \times D^2 \times K}{L}$, which applies to the case—see illustration No. 3 —and take a coefficient of 46 cwts., as given in a well-known pocket-book, the bar would break under a load of about $20\frac{1}{2}$ cwts. only. Working by the rule given in another widely-read and useful pocket-book the bar would break under a load of 24 cwts.

Some engineers, instead of taking BD^2 and using a small coefficient, prefer to take the useful modulus of the section, which is considered to be $\frac{1}{6} BD^2$, and to use a coefficient six times greater than when using BD^2. In the case of the test bar referred to, if the modulus of the section—viz., $\frac{1}{6} BD^2$—is taken, the coefficient of rupture, assuming the bar breaks with 30 cwts., is just over 20 tons. If the bar breaks with 40 cwts., the coefficient of rupture is about 27 tons. These coefficients are much higher than the ultimate tensile stress of cast iron, and the explanation is to be sought in the plasticity of the metal. Sir Benjamin Baker found that if a beam loaded in the centre was turned over and over, it would break after a very few turns under a load, which, under ordinary circumstances, the beam would carry indefinitely.

The arrangement of balls in the illustrations shows how the weight is placed—*i.e.*, whether evenly distributed or placed in the centre of the beam. In Nos. 5 and 6 the ends of the beam are built into the walls or otherwise fixed; this renders the beam stronger than if merely supported, as in the case of Nos. 4 and 5.

The following examples will doubtless help to make the formulæ clear:—

Example 1.—The cast-steel arm of the large hydraulic riveter shown by Fig. 90 is 12 feet long (12 feet gap) and is 45 inches deep from centre to centre of flanges. A pressure of 100 tons is applied at the end ; what stress will there be in the flanges? The formula $\frac{WL}{D}$ given in column 1, opposite the first illustration, applies, and the calculation is $\frac{100 \times 144}{45} = 320$. The stress in each flange is, therefore, 320 tons. The flange is 24 inches wide and 6 inches thick, so that there are 144 square inches to withstand the stress; the stress upon the metal will, therefore, be 2·22 tons per square inch.

Example 2.—A rolled steel joist of I section 6 inches wide and 10 inches deep (say 9½ inches from centre to centre of flange) is built securely into walls 10 feet apart, and is uniformly loaded with a total weight of 26·2 tons; what stress will there be in each flange? The formula given in column 1, opposite the sixth illustration, applies, and the calculation is $\frac{26 \cdot 2 \times 120}{12 \times 9 \cdot 5} = 27 \cdot 6$ tons in each flange. We see, from a sectional drawing of the joist supplied by the makers, that there are about 4 square inches in each flange, so that the stress in the flange is 6·9 tons per square inch.

Example 3.—What weight suspended from the centre of a cast-iron bar 2 inches deep 1 inch wide, placed upon supports 36 inches apart, would break it? The formula given in column 2, opposite the third illustration, applies, and, taking the coefficient value of K to be 70 cwts., the calculation will be $\frac{4 \times 1 \times 4 \times 70}{36} = 31 \cdot 1$ cwts.

Example 4.—What weight suspended from the centre of a cast-iron round bar 3 inches diameter, placed upon supports 36 inches apart, would break it? The value of the metal in a round section is given by the eighth illustration and accompanying figures—viz., $4 \cdot 7 \times R^3$. The calculation, therefore, is $\frac{4 \times 4 \cdot 7 \times 1 \cdot 5^3 \times 70}{36} = 123$ cwts.

The following table, issued by Messrs. Measures Bros., gives the safe permanent loads which their standard rolled-steel joists are capable of bearing when the ends are built into walls 10 feet apart, and when the load is evenly distributed :—

Size.	Weight per Foot.	Safe Load, in Tons, for a 10-feet Span.
Inches.	Lbs.	
6 × 5	24½	9·8
8 × 5	27	13·0
8 × 6	30	15·0
9½ × 4½	24	14·0
10 × 4½	27½	17·0
10 × 5	30	18·6
10 × 6	42	26·2
12 × 5	37	27·5
12 × 6	45	33·2
14 × 6	48	44·1
16 × 6	62	63·0

As the strength of a beam varies inversely as its length, it is easy to calculate what a beam of a length other than 10 feet will carry; thus, a beam 20 feet between walls will carry half the weight given above. If the ends of the beam are merely supported, and not built into the walls, the beam will only carry two-thirds of the weight given above (or in the ratio of 8 to 12). The ratio of strengths of beams, variously loaded and supported, will be seen by glancing at the illustrations and the formulæ in columns 1 and 2.

In connection with structural ironwork, the student may occasionally find it useful to be able to ascertain the stress in any part of a structure by the graphic method. Fig. 104 shows this method applied to a jib crane having a weight suspended from a hook at the end. Assuming the weight suspended from A is 3 tons, and it is desired to know what stress will be set up in the jib and tie bars, the method of procedure is as follows:—
Draw A B equal to the weight; thus, if the weight is 3 tons, the line on a large-scale drawing may be drawn 3 inches long. Draw the horizontal line B C, then C A will show what stress there is in the jib—*i.e.*, if C A is 4 inches long, the stress in the jib is 4 tons. To find the stress in the tie bars, draw the vertical line C D, then D A will show the stress in the tie bars. If the bars had been horizontal, as indicated by dotted lines, the stress in them would have been shown by E A. The stress in the jib and tie bars may also be found by simple arithmetic; thus, the stress in the jib is found by multiplying the weight carried by the length of the jib, and dividing by the distance F G (not F X). The stress in the bars is found by multiplying the weight carried by the length of the bars, and dividing by the distance F G. The student would do well to draw out a triangle to a fairly large scale, take an imaginary weight, and check the graphical method by arithmetical calculations.

The extremely simple methods given above apply only if the weight is suspended from the end of the jib. If the chain carrying the weight passes along the tie bars and is pulled at by a winch, then it is necessary to take the resultant force which arises from the pull on the chain at one end and the weight at the other. This resultant force is found very easily by the graphical method, as shown by Fig. 105. The chain is shown by the dotted line; mark off A B equal to the weight, and mark off A C equal to the pull upon the chain; as the stress in all parts of the chain is the same, it is obvious that A C = A B. Complete the parallelogram C A B D, and the resultant force is shown by the line D A. If now we wish to find the stresses set up in the tie bars and jib of a crane in which the chain passes along the tie bars the method of doing so is shown by Fig. 106. The stress in the tie bars is shown by the line M, and in the jib by the line P.

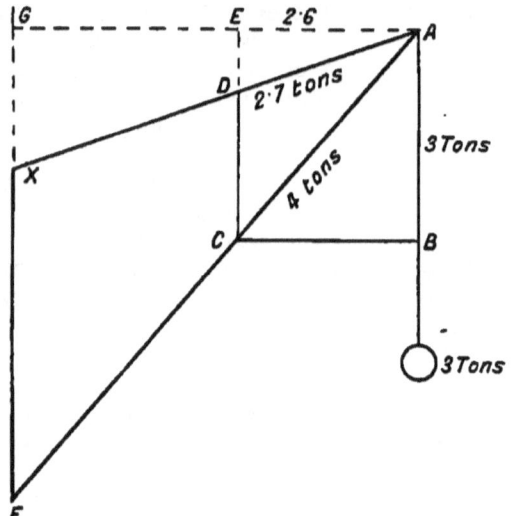

Fig. 104.

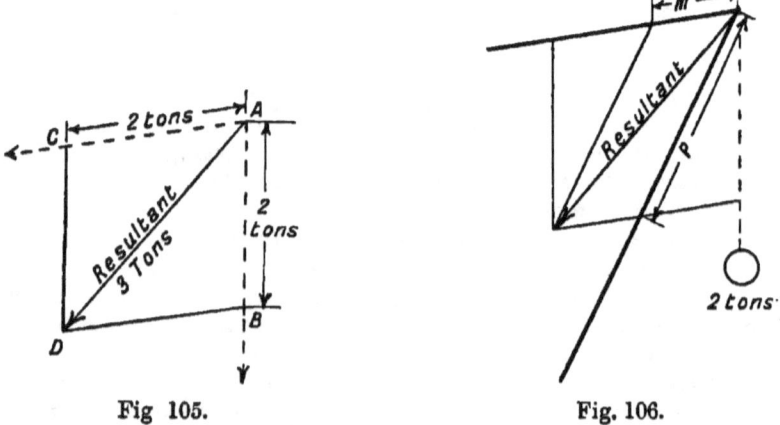

Fig 105.   Fig. 106.

Figs. 104 to 106.—Graphic methods of computing stresses.

A point to be borne in mind in the construction of cranes is this—If the angle formed by the more or less horizontal portion of the chain and the jib is greater than the angle formed by the jib and vertical portion of the chain, then the tie bars will be in compression; if less, then the tie bars will be in tension.

Amongst the wharfs and warehouses of big cities one occasionally sees cranes with the tie bars bent. This is because such cranes were designed by men without the knowledge of the simple fact just mentioned; and small round tie bars, which would be quite suitable for a tensile stress, have been subjected to compression and have consequently bent. The author met one crane-maker who refused to believe that the tie bars of a jib crane could ever be in compression, and it was not until ocular proof was given by means of a lawn-tennis post placed as a jib, and a piece of cord and weight representing the chain, that the crane-maker was convinced. He had attributed the fact of the tie bars of his cranes becoming bent to the practice of men crawling along them to oil the sheave at the end! If the jib is placed at an angle of 45° both in respect to the horizontal and vertical portions of the chain, then the resultant thrust is entirely taken by the jib, and the tie bars are neither in tension nor compression; but, of course, any swing of the chain will upset this equilibrium at once.

The pull on the wall at the top bracket of a crane equals the weight multiplied by the radius of the crane, and divided by the height of the crane post.

Sundry Useful Information.—All questions relating to the power transmitted by a screw, lever, wedge, or toggle joint, can be answered by the following formula :—

$$P = \frac{p\,m}{M}, \qquad p = \frac{P\,M}{m};$$

where P = power transmitted.
p = power applied.
M = motion of power transmitted.
m =   ,,   ,,   applied.

Example.—The hand-wheel of a stop valve is 30 inches in circumference, and, for the sake of a simple example, we will assume that it is keyed to a screwed spindle having only one thread per inch, so that during one revolution of the hand-wheel the valve will be raised or lowered 1 inch. A force of 10 lbs. is exerted at one portion of the hand-wheel, what power will be transmitted to the valve? The power applied will move through 30 inches, while the power transmitted will move through 1 inch. The calculation, therefore, is $\frac{10 \times 30}{1} = 300$. The power transmitted to the valve is 300 lbs.

Questions arising in connection with the use of levers can be answered by the help of the following formula (the formula is transposed for the convenience of beginners):—

$$\frac{W \times y}{x} = P, \qquad \frac{P \times x}{W} = y,$$

$$\frac{W \times y}{P} = x, \qquad \frac{P \times x}{y} = W;$$

where W = weight.
y = its leverage.
P = pressure.
x = its leverage.

Example 1.—A weight of 20 lbs. is placed at the end of a safety-valve lever 12 inches from the fulcrum; the lever acts on the valve at a distance of 1½ inches from the fulcrum; what pressure will be exerted on the valve? The formula is

$$\frac{W \times y}{x} = P, \text{ or } \frac{20 \times 12}{1\cdot 5} = 160 \text{ lbs.}$$

Example 2.—Suppose the pressure thus obtained is not sufficiently great, and we require a pressure of 250 lbs. on the valve, how long must the lever be, keeping the weight the same? The formula transposed to meet this case is

$$\frac{P \times x}{W} = y, \text{ or } \frac{250 \times 1\cdot 5}{20} = 18\cdot 75.$$

The lever must, therefore, be 18·75 inches long.

Thermometer Scales.—The reader will find in the technical press and elsewhere that temperatures are frequently given in degrees Centigrade. To convert degrees Centigrade into degrees Fahrenheit it is necessary to multiply by 9, divide by 5, and add 32 to the result. Thus—

```
     100° C.                                          212° F.
        9                                             - 32
     ─────      To convert degrees Fahren-           ─────
   5 ) 900      heit into degrees Centigrade,         180
     ─────      proceed thus—                           5
      180                                            ─────
     + 32                                          9 ) 900
     ─────                                           ─────
     212° F.                                          100°.
```

The reader will see from the above examples that 100° C. are equivalent to 212° F.

Boiler Pressures.—In France pressures are spoken off in terms of kilogrammes per square centimetre. One kilo. per square centimetre = 14·2 lbs. per square inch, or a little less than one atmosphere. It is useful to remember this, as it enables one to make a rough mental calculation as to the pressure in pounds per square inch, when it is referred to in the other and less familiar way.

To Divide a Straight Line into a Number of Equal Parts.—It is frequently necessary to divide a line of some uneven length into a number of equal parts. For instance. let us suppose that the line A B = $2\frac{15}{16}$ inches long, and we wish to

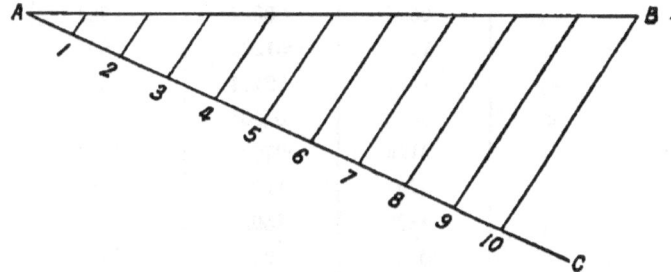

divide it into 10 equal parts. The best way to proceed is as follows:—

Draw A C at any convenient angle and of any length. Divide this line into 10 parts of equal length; it does not matter what the length is so long as the divisions are all equal. Draw a line from the 10th point to B, and lines from the other points of division parallel to 10 B. A B will then be divided by these lines into ten equal parts.

TABLE XXII.—Decimal Equivalents of an Inch with Areas and Circumferences.

Fraction of an Inch.	Decimals of an Inch.	Area.	Circumference.
1/16	·0625	·00307	·1963
1/8	·125	·01227	·3927
3/16	·1875	·02761	·589
1/4	·25	·04909	·7854
5/16	·3125	·0767	·9817
3/8	·375	·1104	1·1781
7/16	·4375	·1503	1·3744
1/2	·5	·1963	1·5708
9/16	·5625	·2485	1·771
5/8	·625	·3068	1·9635
11/16	·6875	·3712	2·1598
3/4	·75	·4417	2·3562
13/16	·8125	·5185	2·5525
7/8	·875	·6013	2·7489
15/16	·9375	·6903	2·9452
1	1·0	·7854	3·1416

Inches.	Decimals of a Foot.	Inches.	Decimals of a Foot.
1	·0833	7	·5833
2	·1667	8	·6667
3	·25	9	·75
4	·3333	10	·8333
5	·4167	11	·9167
6	·5	12	1·0

STRENGTH OF BEAMS AND USEFUL INFORMATION.

TABLE XXIII.—AREAS AND CIRCUMFERENCES OF CIRCLES ADVANCING BY EIGHTHS.

Diameter	0 Area	0 Circumference	1/8 Area	1/8 Circumference	1/4 Area	1/4 Circumference	3/8 Area	3/8 Circumference	1/2 Area	1/2 Circumference	5/8 Area	5/8 Circumference	3/4 Area	3/4 Circumference	7/8 Area	7/8 Circumference
0	0	0	·012	·392	·049	·785	·11	1·178	·196	1·57	·306	1·963	·441	2·356	·601	2·748
1	·785	3·141	·994	3·534	1·227	3·927	1·484	4·319	1·767	4·712	2·073	5·105	2·405	5·497	2·761	5·89
2	3·141	6·283	3·546	6·675	3·976	7·068	4·43	7·461	4·908	7·854	5·411	8·246	5·930	8·639	6·491	9·032
3	7·068	9·424	7·669	9·817	8·295	10·21	8·946	10·602	9·621	10·995	10·32	11·388	11·04	11·781	11·79	12·173
4	12·56	12·566	13·36	12·959	14·18	13·351	15·03	13·744	15·9	14·137	16·8	14·529	17·72	14·922	18·66	15·315
5	19·63	15·708	20·63	16·1	21·64	16·493	22·69	16·886	23·75	17·278	24·85	17·671	25·96	18·064	27·1	18·456
6	28·27	18·849	29·46	19·242	30·67	19·635	31·71	20·027	33·18	20·42	34·47	20·813	35·78	21·205	37·12	21·598
7	38·48	21·991	39·87	22·383	41·28	22·776	42·71	23·169	44·17	23·562	46·06	23·954	47·17	24·347	48·7	24·74
8	50·26	25·132	51·84	25·525	53·45	25·918	55·08	26·31	56·74	26·703	58·42	27·096	60·13	27·480	61·86	27·881
9	63·61	28·274	65·39	28·667	67·2	29·059	69·02	29·452	70·88	29·845	72·75	30·237	74·66	30·63	76·58	31·023
10	78·54	31·416	80·51	31·808	82·51	32·201	84·55	32·494	86·59	32·986	88·66	33·379	90·76	33·772	92·88	34·164
11	95·03	34·557	97·2	34·95	99·5	35·343	101·62	35·735	103·86	36·128	106·13	36·521	108·43	36·913	110·76	37·306
12	113·09	37·699	115·46	38·091	117·85	38·484	120·27	38·877	122·71	39·27	125·18	39·662	127·67	40·055	130·19	40·448
13	132·73	40·84	135·29	41·233	137·88	41·626	140·5	42·018	143·13	42·411	145·8	42·804	148·49	43·197	151·2	43·589
14	153·93	43·982	156·7	44·375	159·48	44·767	162·29	45·16	166·13	45·553	167·99	45·945	170·87	46·338	173·78	46·731
15	176·71	47·124	179·67	47·516	182·65	47·909	185·66	48·302	188·69	48·694	191·74	49·087	194·82	49·48	197·93	49·827
16	201·06	50·265	204·21	50·658	207·39	51·051	210·59	51·443	213·82	51·836	217·07	52·229	220·35	52·621	223·65	53·014
17	226·98	53·407	230·33	53·799	233·70	54·192	237·1	54·585	240·52	54·978	243·97	55·37	247·45	55·763	250·94	56·156
18	254·47	56·548	258·01	56·941	261·58	57·334	265·18	57·726	268·8	58·119	272·44	58·512	276·11	58·905	279·81	59·297
19	283·52	59·69	287·27	60·083	291·04	60·475	294·83	60·868	298·64	61·261	302·48	61·653	306·35	62·046	310·24	62·439
20	314·16	62·832	318·09	63·224	322·06	63·617	326·06	64·01	330·06	64·402	334·1	64·795	338·16	65·188	342·25	65·58
21	346·36	65·973	350·49	66·366	354·65	66·759	358·84	67·151	363·05	67·544	367·28	67·937	371·54	68·329	375·82	68·722
22	380·13	69·115	384·46	69·507	388·82	69·9	393·2	70·293	397·6	70·686	402·03	71·078	406·49	71·471	410·97	71·864
23	415·47	72·256	420	72·639	424·55	73·042	429·13	73·434	433·73	73·827	438·36	74·22	443·01	74·613	447·69	75·005
24	452·39	75·398	458·11	75·791	461·86	76·183	466·63	76·576	471·43	76·969	476·25	77·361	481·1	77·754	485·97	78·147
25	490·87	78·54	495·79	78·932	500·74	79·325	505·81	79·718	510·7	80·11	515·72	80·503	520·76	80·896	525·83	81·288

TABLE XXIII.—Continued.

Diameter.	0		1/8		1/4		3/8		1/2		5/8		3/4		7/8	
	Area.	Circumference.	Area.	Circumference.	Area.	Circumference.	Area.	Circumference.	Area.	Circumference.	Area.	Circumference.	Area.	Circumference.	Area.	Circumference.
26	530·93	81·681	530·04	82·074	541·19	82·467	540·35	82·859	551·14	83·252	556·76	83·645	562	84·037	507·26	84·43
27	572·55	84·823	577·87	85·215	583·2	85·608	588·57	86·001	593·95	86·394	599·37	86·786	604·8	87·179	610·26	87·572
28	615·75	87·964	621·26	88·357	626·79	88·75	632·35	89·142	637·94	89·535	643·54	89·928	649·18	90·321	654·84	90·713
29	660·52	91·106	666·22	91·490	671·95	91·891	677·71	92·284	683·49	92·677	689·29	93·069	695·12	93·462	700·98	93·855
30	706·86	94·248	712·76	94·64	718·69	95·033	724·64	95·426	730·61	95·818	736·61	96·211	742·64	96·604	748·69	96·996
31	754·79	97·389	775·80	95·782	760·99	98·175	773·14	98·56	779·31	98·96	785·51	99·35	791·73	99·745	797·97	100·13
32	804·25	100·53	810·54	100·923	816·86	101·310	823·21	101·70	829·57	102·12	835·97	102·49	842·39	102·887	848·83	103·28
33	855·3	103·07	861·79	104·065	868·3	104·459	874·85	104·85	881·41	105·24	888	105·63	894·62	106·029	901·25	106·42
34	907·92	106·81	914·61	107·207	921·32	107·6	928·00	107·99	934·82	108·38	941·6	108·77	948·42	109·17	955·25	109·56
35	962·11	109·95	969	110·349	975·9	110·74	982·84	111·13	989·8	111·52	996·78	111·91	1003·7	112·31	1010·8	112·70
36	1017·8	113·09	1024·9	113·49	1032·0	113·88	1039·1	114·27	1046·3	114·66	1053·5	115·06	1060·7	115·45	1067·9	115·84
37	1075·2	116·23	1082·4	116·63	1089·7	117·02	1097·1	117·41	1104·4	117·81	1111·8	118·20	1119·2	118·59	1126·6	118·98
38	1134·1	119·38	1141·5	119·77	1149·0	120·16	1156·6	120·55	1164·1	120·95	1171·7	121·34	1179·3	121·73	1186·9	122·13
39	1194·5	122·52	1202·2	122·91	1209·9	123·30	1217·6	123·7	1225·4	124·09	1233·1	124·48	1240·9	124·87	1248·7	125·27
40	1256·6	125·66	1264·5	126·05	1272·4	126·44	1280·4	126·84	1288·2	127·23	1296·2	127·62	1304·2	128·02	1312·2	128·41
41	1320·2	128·80	1328·3	129·19	1336·5	129·59	1344·5	129·98	1352·6	130·37	1360·8	130·76	1369	131·16	1377·2	131·55
42	1385·4	131·94	1393·7	132·34	1401·9	132·73	1410·3	133·12	1418·6	133·51	1426·9	133·91	1435·3	134·30	1443·7	134·69
43	1452·2	135·08	1460·6	135·48	1469·1	135·87	1477·6	136·26	1486·1	136·66	1494·7	137·05	1503·3	137·44	1511·9	137·83
44	1520·5	138·23	1529·1	138·62	1537·8	139·01	1546·5	139·40	1555·2	139·80	1564·1	140·19	1572·8	140·58	1581·6	140·07
45	1590·4	141·37	1599·2	141·76	1608·1	142·15	1617·0	142·55	1625·9	142·94	1634·9	143·33	1643·8	143·72	1652·8	144·12
46	1661·0	144·51	1670·9	144·90	1680·0	145·29	1689·1	145·692	1698·2	146·08	1707·3	146·47	1716·5	146·87	1725·7	147·26
47	1734·9	147·65	1744·1	148·04	1753·4	148·44	1702·7	148·833	1772·0	149·22	1781·4	149·61	1790·7	150·01	1800·1	150·40
48	1809·5	150·79	1819	151·18	1828·4	151·58	1837·9	151·975	1847·4	152·30	1856·9	152·76	1806·5	153·15	1876·1	153·54
49	1885·7	153·93	1895·3	154·33	1905·0	154·72	1914·7	155·116	1924·4	155·50	1934·1	155·90	1943·9	156·29	1953·6	156·08

TABLE XXIII.—Continued.—AREAS AND CIRCUMFERENCES OF CIRCLES.

Area		Circumference		Area		Circumference		Area		Circumference		Area		Circumference
1,963	50	157·1		4,417	75	235·6		7,854	100	314·1		12,277	125	392·7
2,042	51	160·2		4,536	76	238·8		8,012	101	317·3		12,469	126	395·8
2,123	52	163·4		4,656	77	241·9		8,171	102	320·4		12,667	127	398·9
2,206	53	166·6		4,778	78	245·0		8,332	103	323·5		12,867	128	402·1
2,290	54	169·6		4,901	79	248·2		8,494	104	326·7		13,069	129	405·2
2,375	55	172·8		5,026	80	251·3		8,659	105	329·8		13,273	130	408·4
2,463	56	175·9		5,153	81	254·5		8,824	106	333·0		13,478	131	411·5
2,551	57	179·1		5,281	82	257·6		8,992	107	336·1		13,684	132	414·6
2,642	58	182·2		5,410	83	260·8		9,160	108	339·2		13,892	133	417·8
2,733	59	185·4		5,541	84	263·9		9,331	109	342·4		14,102	134	420·9
3,019	60	188·5		5,674	85	267·0		9,503	110	345·5		14,313	135	424·1
3,022	61	191·6		5,808	86	270·2		9,676	111	348·7		14,526	136	427·2
3,019	62	194·8		5,944	87	273·3		9,852	112	351·8		14,741	137	430·3
3,117	63	197·9		6,082	88	276·5		10,028	113	355·0		14,957	138	433·5
3,217	64	201·1		6,221	89	279·6		10,207	114	358·1		15,174	139	436·6
3,318	65	204·2		6,361	90	282·8		10,386	115	361·1		15,393	140	439·8
3,421	66	207·3		6,503	91	285·9		10,568	116	364·4		15,614	141	442·9
3,525	67	210·5		6,647	92	289·0		10,751	117	367·5		15,836	142	446·1
3,631	68	213·6		6,792	93	292·2		10,935	118	370·7		16,060	143	449·2
3,739	69	216·8		6,939	94	295·3		11,220	119	373·8		16,286	144	452·3
3,848	70	219·9		7,088	95	298·5		11,309	120	376·9		16,530	145	455·4
3,959	71	223·1		7,238	96	301·6		11,499	121	380·1		16,741	146	458·0
4,071	72	226·2		7,389	97	304·7		11,689	122	383·2		16,971	147	461·8
4,185	73	229·3		7,542	98	307·9		11,882	123	386·4		17,203	148	464·9
4,300	74	232·9		7,697	99	311·0		12,076	124	389·5		17,436	149	468·0
17,671	150	471·2		24,025	175	549·7
17,907	151	474·3		24,328	176	552·9
18,145	152	477·5		24,60 [	177	556·0
18,385	153	480·6		24,884	178	559·2
18,626	154	483·8		25,165	179	562·3
18,369	155	486·9		25,446	180	565·4
19,113	156	490·0		25,730	181	568·6
19,359	157	493·2		26,015	182	571·1
19,607	158	496·3		26,302	183	574·9
19,855	159	499·5		26,590	184	578·0
20,106	160	502·6		26,880	185	581·1
20,358	161	505·7		27,171	186	584·3
20,612	162	508·9		27,464	187	587·4
20,867	163	512·0		27,759	188	590·6
21,124	164	515·2		28,055	189	593·7
21,382	165	518·3		28,352	190	596·9
21,642	166	521·5		28,652	191	600·0
21,904	167	524·6		28,952	192	603·1
22,167	168	527·8		29,253	193	606·3
22,431	169	530·9		29,559	194	609·4
22,698	170	534·0		29,864	195	612·6
22,965	171	537·2		30,171	196	615·7
23,235	172	540·3		30,480	197	618·8
23,506	173	543·5		30,790	198	622·0
23,778	174	546·5		31,102	199	625·1
				31,416	200	628·3

267

TABLE XXIV.—TABLE OF SQUARES AND CUBES.

No.	Square.	Cube.	No.	Square.	Cube.	No.	Square.	Cube.	No.	Square.	Cube.
1	1	1	51	2,601	132,651	101	10,201	1,030,301	151	22,801	3,442,951
2	4	8	52	2,704	140,608	102	10,404	1,061,208	152	23,104	3,511,808
3	9	27	53	2,809	148,877	103	10,609	1,092,727	153	23,409	3,581,577
4	16	64	54	2,916	157,464	104	10,816	1,124,864	154	23,716	3,652,264
5	25	125	55	3,025	166,375	105	11,025	1,157,625	155	24,025	3,723,875
6	36	216	56	3,136	175,616	106	11,236	1,191,016	156	24,336	3,796,416
7	49	343	57	3,249	185,193	107	11,449	1,225,043	157	24,649	3,869,893
8	64	512	58	3,364	195,112	108	11,664	1,259,712	158	24,964	3,944,312
9	81	729	59	3,481	205,379	109	11,881	1,295,029	159	25,281	4,019,679
10	100	1,000	60	3,600	216,000	110	12,100	1,331,000	160	25,600	4,096,000
11	121	1,331	61	3,721	226,981	111	12,321	1,367,631	165	27,220	4,492,120
12	144	1,728	62	3,844	238,328	112	12,544	1,404,928	170	28,900	4,913,000
13	169	2,197	63	3,969	250,047	113	12,769	1,442,897	175	30,620	5,359,370
14	196	2,744	64	4,096	262,144	114	12,996	1,481,544	180	32,400	5,832,000
15	225	3,375	65	4,225	274,625	115	13,225	1,520,875	185	34,220	6,331,600
16	256	4,096	66	4,356	287,496	116	13,456	1,560,896	190	36,100	6,859,000
17	289	4,913	67	4,489	300,763	117	13,689	1,601,613	195	38,020	7,414,000
18	324	5,832	68	4,624	314,432	118	13,924	1,643,032	200	40,000	8,000,000
19	361	6,859	69	4,761	328,509	119	14,161	1,685,159	210	44,100	9,261,00
20	400	8,000	70	4,900	343,000	120	14,400	1,728,000	220	48,400	10,684,000
21	441	9,261	71	5,041	357,911	121	14,641	1,771,561	230	52,900	12,167,000
22	484	10,648	72	5,184	373,248	122	14,884	1,815,848	240	57,600	13,824,000
23	529	12,167	73	5,329	389,017	123	15,129	1,860,867	250	62,500	15,625,000
24	576	13,824	74	5,476	405,224	124	15,376	1,906,624	260	67,900	17,576,000
25	625	15,625	75	5 625	421,875	125	15,625	1,953,125	270	72,900	19,683,000
26	676	17,576	76	5,776	438,976	126	15,876	2,000,376	280	74,840	21,952,000
27	729	19,683	77	5,929	456,533	127	16,129	2,048,383	290	84,100	24,389,000
28	784	21,952	78	6,084	474,552	128	16,384	2,097,152	300	90,000	27,000,000
29	841	24,389	79	6,241	493,039	129	16,641	2,146,689	310	96,100	29,791,000
30	900	27,000	80	6,400	512,000	130	16,900	2,197,000	320	102,400	32,768,000
31	961	29,791	81	6,561	531,441	131	17,161	2,248,091	330	108,900	35,937,000
32	1,024	32,768	82	6,724	551,368	132	17,424	2,299,968	340	115,600	39,340,000
33	1,089	35,937	83	6,889	571,787	133	17,689	2,352,637	350	122,500	42,575,000
34	1,156	39,304	84	7,056	592,704	134	17,956	2,406,104	360	129,600	46,656,000
35	1,225	42,875	85	7,225	614,125	135	18,225	2,460,375	370	136,900	50,653,000
36	1,296	46,656	86	7,396	636,056	136	18,496	2,515,456	380	144,400	54,872,000
37	1,369	50,653	87	7,569	658,503	137	18,769	2,571,353	390	152,100	59,319,000
38	1,444	54,872	88	7,744	681,472	138	19,044	2,628,072	400	160,000	64,000,000
39	1,521	59,319	89	7,921	704,969	139	19,321	2,685,619	410	168,100	68,921,000
40	1,600	64,000	90	8,100	729,000	140	19,600	2,744,000	420	176,400	74,088,000
41	1,681	68,921	91	8 281	753,571	141	19,881	2,803,221	430	184,900	79,507,000
42	1,764	74,088	92	8,464	778,688	142	20,164	2,863,288	440	193,600	85,184,000
43	1,849	79,507	93	8,649	804,357	143	20,449	2,924,210	450	202,500	91,125,000
44	1,936	85,184	94	8,836	830,584	144	20,736	2,985,980	460	211,600	97,336,000
45	2,025	91,125	95	9,025	857,375	145	21,025	3,048,620	470	220,900	103,823,000
46	2,116	97,336	96	9,216	884,736	146	21,316	3,112,136	480	230,400	110,572,000
47	2,209	103,823	97	9,409	912,673	147	21,609	3,176,523	490	240,100	117,649,000
48	2,304	110,592	98	9,604	941,192	148	21,904	3,241,792	500	250,000	125,000,000
49	2,401	117,649	99	9,801	970,299	149	22,201	3,307,949	600	360,000	216,000,000
50	2,500	125,000	100	10,000	1,000,000	150	22,500	3,375,000	700	490,000	343,000,000

WEIGHTS AND MEASURES WITH METRIC EQUIVALENTS.

LENGTH.

```
                1 inch   =   25·399 millimetres.
   12 inches  = 1 foot   =  304·799      ,,
    3 feet    = 1 yard   =  914·391      ,,
                   ,,    =     ·914 metre.
1,760 yards   = 1 mile   =    1·609 kilometres.

              1 metre      =  39·3704 inches.
              1 decimetre  =   3·9370    ,,
              1 centimetre =    ·3937 inch.
              1 millimetre =    ·0393    ,,
              1 kilometre  = 1,093 yards.
```

WEIGHT.

```
                    1 ounce                       =   28·35 grams.
16 ounces         = 1 pound                       =  453·59    ,,
                    1   ,,                        =    ·453 kilogram.
28 pounds         = 1 quarter                     =   12·7  kilograms.
 4 quarters       = 1 hundredweight (112 lbs.)   =   50·8     ,,
20 hundredweights = 1 ton (2,240 lbs.)           = 1,016      ,,

              1 kilogram = 2·2046 lbs.
              1 gram     =  ·0022 lb.
```

AREA.

```
                 1 sq. inch  = 645      sq. millimetres.
144 sq. inches = 1 sq. foot  =   ·0929 sq. metre.
  9 sq. feet   = 1 sq. yard  =    ·836    ,,
4,840 sq. yards = 1 acre     =   ·4046 Hectare.

              1 sq. metre = 10·764 feet.
```

VOLUME.

```
                  1 cubic inch = 16·387 cubic centimetres.
1,728 cubic inches = 1  ,,   foot  =   ·0283  ,,  metre.
   27    ,,   feet = 1  ,,   yard  =   ·7645  ,,    ,,

              1 cubic metre     =  35·3148 cubic feet.
              1   ,,    ,,      =   1·3079   ,,   yards.
              1   ,,  centimetre =    ·061   ,,  inch.
              1 gallon          =    ·1605   ,,  foot.
              1   ,,            =  277·27    ,,  inches.
              1   ,,            =    4·54  litres.
```

CONCLUSION.

A few words of advice from one who has been through the mill may perhaps not be taken amiss.

In the first place, a man who has "got it all down in a book at home" is not of very much use in a drawing office, as questions usually require to be settled when they arise. Secondly, it is quite inadmissible to have text-books lying around one's drawing board. Engineering pocket-books containing tables of areas, squares, &c., are, however, permitted, and as they usually contain some blank pages at the end, it is a good plan to copy into them any formulæ or rules which the student thinks may be likely to be useful to him. If the beginner has not got one of these pocket-books, then he should copy into a small note-book of his own any useful information which he may come across. Such a note-book, to be of any real use, should be small and thin, so that it may be carried in the pocket without inconvenience.

Amongst the information which a young draughtsman should have available at a moment's notice, the following may be mentioned :—Power transmitted by belts, ropes, and shafts; rules as to size of steam pipes; areas of chimneys; head necessary to overcome friction of water in pipes; stress set up in the rim of a flywheel due to centrifugal force; stresses in beams, &c. A young draughtsman who can settle questions such as these, at all events in connection with preliminary drawings such as are frequently sent out, will have a much better chance of promotion than one who is continually troubling his superior with questions such as, "How wide must I make the pulley?" "What size must the steam pipe be?" "How large a chimney shall I show?" &c., &c.

The next question which arises is—How much mathematical knowledge is it necessary for a young engineer to have in order to succeed? The reply is—The more knowledge one has, the better one is equipped for one's work; but if, in the acquirement of such knowledge, one loses other qualifications for success, such as health, good eyesight, strength of character, and robust commonsense, then the acquirement of profound mathematical knowledge may be too dearly bought.

The brain of a highly-trained mathematical man may not unfitly be compared to a razor, while that of the ordinarily successful engineer and man of business to an axe, which is not easily turned aside by opposition. If an individual is compelled by force of circumstances to go out to the backwoods, make his own clearing, and build his own house, there can be little doubt which of the two instruments will prove of most service. The author has no desire to decry the razor, but, at the same time, he would not recommend a young man with his own clearing to make to attempt to grind and temper his comparatively blunt, but serviceable, instrument until it had the edge of a razor, but at the expense of other and more valuable qualities.

The late Lord Thring stated he "found that the apparent object of legal expression was to conceal the meaning from ordinary readers," and although the author has great respect for the learning of Professors of Engineering, he cannot help feeling that the methods of many of them, while not designed to conceal their meaning, yet are calculated to deter the ordinary youth from studying the theoretical side of engineering at all.

Take, for instance, the following statement which appears in a book intended for young students of engineering:—"In single shear, the shearing resistance of the rivet is $\frac{\pi}{4} d^2 f'$; where f' is the resistance of the material to shearing." Would it not have been simpler to state that in single shear the shearing resistance of the rivet is $A \times R$; where A = area, R = resistance of the metal to shearing? An expression such as $\frac{\pi}{4} d^2 f'$ is all very well for professors and others who can see at a glance that $\frac{\pi}{4} d^2$ stands for area; but if a conscientious beginner laboriously squares the diameter of the rivet, multiplies the result by 3·141, then divides by 4, only to find that the whole of this labour conveys the remarkable intelligence that the strength of a rivet equals its area multiplied by the ability of the metal to resist shear, he will probably say that his own commonsense would tell him that, and will struggle no further with other formulæ which might really be of use. If the beginner thinks any more about the subject, he will wonder why in the name of common-sense those gentlemen who are so fond of using Greek symbols at every possible opportunity, do not choose one to represent the area of a circle, and stick to it, instead of writing $\frac{\pi}{4} d^2$.

Again, if a student should ask a junior professor how to find the pressure on the slide bars of an engine, he will probably be told that the pressure can be found by multiplying the pressure on the piston by the sine of the angle formed by a prolongation of the piston-rod and connecting-rod! Now, in a drawing office where draughtsmen's time has to be paid for, one does not find them seeking out angles by means of protractors, and then turning up tables of sines, cosines, secants, &c. If a draughtsman wishes to know what the greatest pressure will be upon the slide bars, he simply multiplies the pressure exerted by the piston by the length of crank, and divides by the length of connecting-rod, ignoring sines and angles altogether.

As an example, showing the effect of a beginner trying to work out simple problems by advanced geometrical methods, of which he has only an imperfect grip, a short narrative of what occurred in a Westminster drawing office, in which the author worked as a junior, may not be out of place. In the office in question, it was necessary frequently to run out the stress set up in riveter arms. So far as the author was privileged to see, no calculations more elaborate than those described in the preceding chapter were ever made, either by the chief draughtsman or his assistants. In fact, a tradition was current in the office that the riveter arms had in the first instance been designed by a gentleman (not the inventor) on ultra-scientific lines, and had all broken. However, whether this tradition is or is not true, it is beside the story, which is this:—There came to the drawing office, as a premium pupil, a young man who had had the benefit of some training at a technical college. This pupil looked upon the ordinary office methods of arriving at the stress set up in a riveter arm with considerable scorn. He said they were quite unscientific, and that he would show the correct way. The next time a riveter of extra large gap, or extra heavy power, was required, the problem was given to the pupil, as well as to one of the older men, to work out. In the case of the pupil, the result was as follows:—After some hours of steady work a diagram was produced, which to the other juniors appeared to be like a small church with its steeple. For the success of the calculations, apparently, it was necessary to find with great accuracy the centre of gravity of the diagram. The little church was, therefore, cut out in cardboard, and attempts made to find the centre of gravity by balancing it on the point of a divider. Unfortunately, just as the desired centre was nearly found, the office door would be opened, a current of air would cause the diagram to fall to the ground, and operations had to be recommenced. However, the centre of gravity having at last been found, more steady work was put in, a book of logarithms was freely used, several sheets of foolscap were covered with figures, and at last the pupil said his calculation showed that the arm would require to be (as nearly as the author can remember) about 5,283 feet deep! When it was pointed out that a riveter arm over a mile across would be rather unusual, the pupil remarked, quite unabashed, that he supposed he had got wrong with his logarithms. The strangest feature about this pupil was that, although he obtained contradictory results every time he worked out the same problem, yet he remained convinced that his was not only the right and scientific way of working, but also the best way. Possibly it was for one who really understood it, and had the time and ability to work out the problem correctly, but the method in question was quite unsuitable for every-day use in a drawing office.

If the student does not take kindly to advanced mathematics he need not be discouraged; only those who have actually worked in drawing offices of good and successful firms know how small a part mathematics play in the daily life of the office. The problem is rather how to get out the work quickly with a limited staff and to avoid mistakes, than to work out elaborate calculations.

The fundamental difference between work in the drawing office of a technical college and work done in a real drawing office, is that a mistake made in the former does not really matter and may possibly pass unnoticed, while a mistake made in the latter may involve a heavy pecuniary loss, or even give rise to a Coroner's inquest. In the drawing office of a technical college the omission to mark the revolving part of a certain machine "cast steel," leaving it to be inferred that it was to be of cast iron, might result in the loss of a few marks. In a Midland town recently, such an omission on the part of a draughtsman caused the loss of a couple of lives. Had the draughtsman been accustomed to run out by simple arithmetic the stresses set up in a flywheel rim, as described in a previous chapter, this accident and the resulting inquest, verdict, and somewhat disagreeable rider, might have been avoided. Unfortunately many draughtsmen are deterred from making calculations, which, although really simple, are made to appear difficult and complicated by so many text-books.

If the young engineer should eventually have a business of his own, his problems will be, not how to make elaborate mathematical calculations, but how to get remunerative work, and, when obtained, how to turn it out quickly and well. In many cases another problem will be added—viz., where to find the money for the following week's wages for his men. If a knowledge of the Calculus would enable a young engineer to solve this problem satisfactorily upon all occasions, there is but little doubt that it would be studied much more widely than is at present the case.

It must be borne in mind that, valuable as mathematical

investigations are in enabling one to understand a subject thoroughly, they usually follow, and do not precede, useful inventions; for instance, had not such comparatively "ignorant" men as Watt and George Stephenson invented and perfected the steam engine, many of our modern Professors, who write so learnedly and brilliantly on the subject, would have been compelled to devote their abilities to astronomy or kindred subjects. Again, let us take the case of the problem of aerial navigation, which, at the moment of writing these pages, has not been solved. Books showing mathematically and theoretically how such flight may be accomplished seem to be strangely lacking, but as soon as some possibly ignorant and benighted inventor constructs an air ship which can be used with safety, then, undoubtedly, the technical press will teem with books filled from cover to cover with mathematics showing exactly how it is done, and the student will be told that, if he wishes to know anything about aerial flight, he must master the contents of these works. One cannot help feeling, however unreasonable the feeling may be, that such books would be still more useful if they were published before, rather than after, the problem has been solved on other lines.

The author has often been struck with the fact that when an engineering firm gets into a slight difficulty, say the ropes will persist in jumping off a certain pulley, or a barometric condenser will not give a vacuum, or possibly a boiler isolating valve knocks itself to pieces, it is not the man with the greatest theoretical and mathematical knowledge who gets over the trouble, but in nearly every case it is some one having the irreducible minimum of mathematical knowledge. The reason for the success of the one man and the failure of the other is probably due to the fact, not that the one man is ignorant of mathematics, but that he can think and reason upon the facts before him undisturbed by the noise of machinery and uncomfortable surroundings, while the other can only reason and think in symbols, and when seated in a quiet room with a clean sheet of paper before him.

In the author's opinion, there is one qualification for success which is of considerably more importance than a knowledge of advanced mathematics; it is, to put it colloquially, the ability to "keep on keeping on." It was entirely by his determination and strength of purpose that the late Mr. Tweddell was able to make the world see the value of his hydraulic riveter, and to reap the financial reward due to him. The same may be said of Mr. Parsons and his turbine, and of many others.

Competition is getting keener every day, and the struggle for a competence more severe, but there is a good deal of truth in the brave words of a modern writer—"Between aspiration and achievement there is no great gulf fixed, only the faith of concentrated endeavour, only the stern years which must hold fast the burden of a great hope, only the patience strong and meek which is content to bow beneath the fatigue of a long and distant purpose, only these stepping stones and no gulf impassable to human feet divide aspiration from achievement." It is consoling, too, sometimes to reflect that, whether one succeeds or whether one fails, a man cannot do more than his level best. As Epictetus said—"It is not for the actor to choose his part, but it is the actor's duty to play the part that is given to him to the best of his ability, the result being left to the director of the play." And that the crippled philosopher accepted his part cheerfully may be gauged from his words—"For what else can I do, an old man and lame, than sing hymns to God? If I were a nightingale, I would do after the manner of a nightingale; if a swan, after the manner of a swan. But now I am a reasoning creature, and it behoves me to sing the praise of God. This is my task, and this I do, nor, as long as it is granted to me, will I ever abandon the post. And you, too, I summon to join me in the same song."

INDEX.

A

ACCUMULATORS, Electric, 206.
„ Hydraulic, 212.
Adamson rings, 29.
Air pump, Capacity of, 169.
Alloys, 8
Alternating current, Periodicity of, 204.
Alternators, 199.
Aluminium, 8.
„ bronze, 10.
Ampère, 195, 209.
Analyses of gases, 249.
Arc lamps, 197.
Area and circumference of circles, 265.
„ of ports, 137.

B

BABBIT metal, 10.
Babcock boiler, 39.
Balanced slide valve, 93.
Batteries, Primary and secondary, 206.
Beams, Strength of, 255.
Bearings, Pressure on engine, 139.
„ Shafting, 158.
Belleville boiler, 40.
Bellis engine, 111.
Belts, Canvas, 151.
„ Leather, 149.
Bessemer steel, 7.
Blake boiler, 36.
Boiler compositions, 54.
„ Feed-water, 54.
„ flues, 44.
Boilers, 27.
„ Combustion of fuel in, 51.
„ Evaporation of, 49.
„ Galvanic action in, 37.
Boilers, Pitting in, 38.
„ Shell-type, 27, et seq.
„ „ dangers of, 37.
„ Steel plates for, 18.
„ Strength of, 41.
„ Testing, 55.
„ Water-tube, 39, et seq.
Bolts and nuts, 19.
Brake, 118.
Brass, 10.
Brazing, 8.
British thermal unit, 47.
Bronzes, 9.
Browett-Lindley engine, 111.
Brush-Parsons turbine, 181.

C

CALORIFIC value of coals, 50.
Calorimeter, 56.
Cameron pump, 22.
Capacity of pumps, 61.
Carbon in iron, 4, 5.
„ in steel, 5, 6, 7.
Casehardening, 1.
Castings, Effect of cooling on, 11.
Cast iron, 2, 17.
„ steel, 5, 17.
Castle nuts, 20.
Centrifugal force, 131.
„ pumps, 228.
Chimneys, Area of, 52.
„ Draught in, 51.
Chromium, Effect of, on steel, 6.
Coal, Calorific value of, 50.
„ conveyers, 69.
Coefficient of friction, 139.
Combustion, Rate of, 51, 53.
Compound engine, 102.
„ „ Calculation of power of, 115.

Condenser, Barometric, 174.
,, Ejector, 173.
,, Evaporative, 173.
,, Jet, 165.
,, Surface, 167.
Connecting-rods, 138.
Consumption of oil and petrol, 250-254.
,, ,, steam in engines, 103.
Conveyers for coal, 69.
Cooling towers, 174.
Copper, 8.
,, wire, Strength of, 15.
Corliss valve gear, 104.
Cornish boilers, 30, 32.
Corrosion in boilers, 37.
,, of condenser tubes, 171.
Cost of materials, 18.
Crank shafts, Material for, 5.
,, Proportions of, 139.
Cruse superheater, 67.
Curtis turbine, 185.
Cylinder walls, Thickness of, 137.
Cylinders, Iron for, 3.
,, Ratio of, 137.

D

DECIMAL equivalents, 264.
De Laval turbine, 182.
Delta metal, 9.
Diesel oil engine, 241, 243, 254.
Dished boiler ends, 43.
Dowel pins, 20.
Dowson gas plants, 246.
Draught, Induced, 52.
,, Natural, 51.
Double-beat valves, 108.
Drop valves, 108.
Dry-back boiler, 37.
Duplex pump, 57.
Dynamos, 193.

E

ECCENTRIC-RODS, 138.
Economic boiler, 33.
Economisers, 63.
Edwards air pump, 172.
Efficiency of gas engines, 245.
,, ,, steam engines, 144.
Ejector condenser, 173.

Elasticity, Young's modulus of, 16.
Elastic limit, 12.
Electrical units, 196, 209.
Electric current, Production of, 193.
Elongation, Percentage of, 12.
Engines, see under *Steam*, *Gas*, *Petrol*, and *Oil*.
Evaporation of a boiler, Calculation of, 49.
,, ,, Cornish boilers, 33.
,, ,, Lancashire boilers, 32.
Evaporative condenser, 173.
,, power of coal, 50.
Exhaust pipes, 82.
Expansion, Governing by, 134, 136.
,, of pipes, 73.

F

FACTORS of safety, 16.
Fahrenheit degrees, Conversion into Centigrade, 262.
Falls of Foyers turbine, 228.
Fatigue of metals, 15.
Feed pumps, 57, *et seq*.
Feed-water, 54.
,, heaters and economisers, 63.
Filters, Oil, 70.
Flanges, Table of, 80, 81.
Flash-point of oils, 254.
Flow of steam through pipes, 74.
,, ,, water through pipes, 235.
Flues, Strength of boiler, 44.
Flywheel pumps, 60.
Flywheels, 129, 133.
,, Energy stored in, 133.
,, Form of arms of, 11.
,, Stress in, 131.
Forced draught, 52.
Foundation bolts, 21.
"From and at" explained, 49.
Froude water brake, 120.
Fuels, Calorific value of, 50.
Furnace, acid and basic lining, 7.

G

GALLOWAY boiler, 33.
Galvanic action in boilers, 37.
Gas engines, 239.
,, suction plant, 246.

Gas threads, 23.
Gearing, Power transmitted by, 158, 162.
„ Worm, 159.
Governing reciprocating engine by expansion, 134, 136.
„ reciprocating engine by throttling, 134, 136.
„ steam turbines, Curtis, 186.
„ „ intermittent, 179.
„ „ throttling, 181.
Grate area, 52.
Grover washer, 20.
Gunmetal, 8.

H

HANS REYNOLDS chain, 162.
Hardening of steel, 6.
„ „ wrought iron, 1.
Heat lost by radiation from pipes, 78.
„ Mechanical equivalent of, 47.
Heating surface of Cornish boilers, 33.
„ „ „ Lancashire boilers, 32.
Helical wheels, 159.
Helicoid nuts, 20.
Hornsby oil engine, 252.
Horse-power, 96.
Howden's forced draught, 69.
Hydraulic accumulator, 212.
„ jack, 218.
„ jigger, 218.
„ lifts, 217.
„ memoranda, 237.
„ press, 211, 217.
„ ram, 235.
„ riveter, 216.
„ testing, 219.
Hyperbolic curve, 99.

I

IGNITION for gas engines, 241.
Impact tests, 12.
Incandescent lamps, 197.
Indicator diagrams, 124.
Indicators, 122.
Induced draught, 52.

Injectors, 61.
Inward-flow turbine, 221.
Impulse water wheel, 225.
Iron, Cast, 2, 17.
„ Malleable cast, 4.
„ Wrought, 1.
Isolating valves, 88.

J

JACKETING steam engines, 114.
Jockey pulleys, 152.
Joints, Expansion, 73.
„ Lap and butt, 25.
„ Strength of, 45.
Joists, Strength of, 255.
Jonval turbine, 225.
Joule's mechanical equivalent of heat, 47.
Journals, Pressure on, 139.

K

KÖRTING gas engine, 243.

L

LANCASHIRE boiler, 27.
Lap of slide valve, 91.
Lead of slide valve, 92.
Lentz valve gear, 108.
Lewis bolts, 21.
Lighting by electricity, 197.
Lock nuts, 20.
Locomotive boiler, 33.

M

MALLEABLE cast iron, 4.
Manganese bronze, 9, 17.
„ Effect of, on steel, 6.
Marine boiler, 36.
Materials, Cost of, 18.
„ Testing, 11.
Mechanical equivalent of heat, 47.
„ stokers, 67.
Meldrum's forced draught, 69.
Metallic packing, 141.
Mild steel, 2, 5, 17.
Modulus of elasticity, 16.

INDEX.

Mond gas, 249.
Muntz metal, 10.

N

NIAGARA FALLS turbines, 228.
Nickel steel, 5, 8.
Niclausse boiler, 40.
Nuts, 19.

O

OHM, 209.
Oil engines, 252.
,, filters, 70.
Open-hearth steel, 7.
Otto cycle gas engines, 239.
Outward-flow turbines, 221.

P

PACKINGS for steam engines, 140.
Parsons steam turbine, 177.
Petrol, Calorific value of, 254.
,, Consumption of, 254.
,, engines, 252.
Petroleum, Calorific value of, 50.
Phosphor bronze, 9.
Phosphorus in steel, Effect of, 8.
Pig iron, Qualities of, 3.
Pipe flanges, 79-81.
Pipes, Steam, 71.
,, Exhaust, 82.
,, Expansion of, 73.
,, Flow of steam in, 75.
,, ,, ,, water in, 235.
,, Size of, 74.
,, Strength of, 78.
,, Water hammer in, 81.
Piston rings and springs, 142.
,, rods, 138.
,, speeds, 140.
Pitting of boiler plates, 38.
Ports, Area of, 137.
Power, Calculation of engine, 96, 113, 244.
,, Loss of, in transmission, 155.
,, transmission by belts, 149.
,, ,, ,, gearing, 158.
,, ,, ,, ropes, 153.
,, ,, ,, shafts, 156.

Pressure and temperature of steam, 48.
,, on bearings, 139.
Priestman oil engines, 252.
Primary battery, 206.
Producer gas, 247.
Proportions of steam engines, 137.
Pulleys, Convexity of, 151.
,, Distance apart of centres, 152, 154.
,, Fast and loose, 153.
,, Form of arms, 11.
,, Grooves for ropes, 154.
,, Jockey, 152.
Pump, Cameron, 60.
,, capacities, 61, 237.
,, Deane, 60.
,, Duplex, 57.
,, Pulsometer, 233.
,, valves, 252.
,, Weir, 58.
Pumps, Materials for, 235.

Q

QUALITY of metals, 16.

R

RADIATION from pipes, 78.
Ramsbottom rings, 142.
Rateau turbine, 188.
Ratio of engine cylinders, 137.
Reduction of areas, 12.
Reversing gear, 92, 95.
Rivets, 24.
,, Strength of, 45.
Rolled steel joists, 258.
Rolling, effect on strength of steel, &c., 15.
Rope driving, 153.
,, ,, Pulleys for, 154.
Rotary converter, 206.

S

SAFETY valves, 30.
Scavenging gas engine cylinders, 242.
Set screws, 19.
Shafting, Power transmission by, 156.

INDEX.

Siemens-Martin open-hearth steel, 7.
Siemens open-hearth steel, 7.
  ,, producer gas, 246.
Skew wheels, 161.
Slide bars, Pressure on, 139.
  ,, valve, Balanced, 93.
  ,,   ,, D, 91.
  ,,   ,, Piston, 91.
Slip in pumps, 61.
Specifications as to quality, 17.
Specific gravity of oils, 254.
  ,, heat, 176.
Speed of belts, 150.
  ,, ,, gas engines, 246.
  ,, ,, pistons, 140.
  ,, ,, ropes, 153.
  ,, ,, turbines, 181.
  ,, ,, Willans engines, 110.
Squares and cubes, 268.
Steady pins, 20.
Steam domes, 30.
  ,, engine parts, Proportions of, 137.
  ,, engines, 89.
  ,,   ,, Consumption, 103.
  ,,   ,, Efficiency of, 144.
  ,,   ,, Proportions of, 137.
  ,,   ,, Testing, 116.
  ,, Latent heat, 47.
  ,, pipes and valves, 71, et seq.
  ,, Properties of, 46.
  ,, -raising accessories, 57.
  ,, Saturated, 49.
  ,, Specific heat, 176.
  ,, Superheated, 49, 66, 78.
  ,, Table of, 48.
  ,, traps, 83.
  ,, turbines, 177.
Steel, 2, 5, 17.
  ,, Hardening and tempering, 6.
  ,, High-speed tool, 6.
  ,, Self-hardening, 6.
Stirling boiler, 41.
Stokers, Mechanical, 67.
Stone's bronze, 9.
Stop valves, 72, 77, 85, 86.
  ,, ,, Obstruction of, 77.
Strength of beams, 255.
  ,, ,, boilers, 41.
  ,, ,, bolts and nuts, 23.
  ,, ,, cylinders, 78.
  ,, ,, materials, 1-15.
  ,, ,, riveted joints, 45.
Stress and strain, 12.
Studs, 19.

Suction gas plant, 246.
Superheated steam, 49, 78.
Superheaters, 66.
Surface condensers, 167.

T

TEETH of wheels, 159, 163.
Temperature, effect on metals and alloys, 16.
Testing boilers, 55.
  ,, elongation, 12.
  ,, impact, 12.
  ,, machines, 11-13.
  ,, materials, 11; see also Tables I. and II.
  ,, steam engines, 116.
Thermal efficiency, 145, 245.
  ,, storage system, 65.
Thornycroft boiler, 41.
Threads, Whitworth and gas, 22, 23.
Throttling, Governing by, 134, 136.
Transmission of heat, 54.
  ,, ,, power, 149.
Tungsten, Effect of, on steel, 6.
Turbines, Economy of, 190.
  ,, Steam, 177.
  ,, Water, 221.

U

UNITED STATES packing, 141.

V

VACUUM augmenter, 171.
Valve, Balanced, 93.
  ,, Corliss, 104.
  ,, D-form, 89.
  ,, Drop, 108.
  ,, Hopkinson-Ferranti, 87.
  ,, Isolating, 88.
  ,, Lentz, 108.
  ,, Piston, 91.
  ,, rods, 138.
  ,, Stop and gate, 72, 85, 86.
Valves and pipes, 71.
Van der Kerchove engine, 108.
Vertical boilers, 37.
Vibration of engines, 111.
Volt, 195, 209.

W

WATER brake, Froude, 120.
,, hammer, 81.
,, -tube boilers, 39.
,, turbines, 221.
,, Weight of, 237.
,, wheels, 220.
Weir pump, 58.
Welding heat, 1.
Westinghouse turbine, 188.
Wheels, Raw-hide, 161.
,, Toothed, 159.
,, Worm, 159.
White metal, 10.
Whitworth compressed steel, 5.
,, threads, 22.
Willans engine, 108.

Willans-Parsons turbine, 180.
Wöhler's law, 16.
Worm gear, 159.
Worthington pump, 57.
Wrought iron, 1.

Y

YARROW boiler, 41.
Young's modulus of elasticity, 16.

Z

ZEUNER diagram, 145.
Zoelly turbine, 189.

www.ingramcontent.com/pod-product-compliance
Lightning Source LLC
Chambersburg PA
CBHW032054230426
43672CB00009B/1591